CHEMTRAILS

CHEMTRAILS

QUINN SILVER

CONTENTS

Introduction

Imagine gazing up at the clear blue sky, marveling at the intricate patterns left by passing airplanes. What if those trails, often dismissed as mere water vapor, held secrets? What if they were chemical agents, intentionally dispersed for hidden agendas? This is the provocative claim at the heart of the chemtrails conspiracy theory.

The chemtrails theory suggests that governments and other powerful entities are covertly spraying chemicals into the atmosphere for reasons ranging from weather modification to mind control. Despite overwhelming scientific evidence debunking these claims, the theory continues to captivate a dedicated following.

Why does this theory persist? Why do people believe in something so widely disputed by experts? This book aims to delve deep into these questions, exploring the origins, the science, the believers, and the debunkers of the chemtrails conspiracy theory. It's a journey through pseudoscience, psychology, and the influence of modern media.

By examining these threads, we'll uncover not just the narrative of the chemtrails theory but also the broader phenomena of conspiracy thinking and the societal forces that nourish it. Prepare to look at the sky—and the information we consume—through a more discerning lens.

Chapter 1: Introduction to Chemtrails

Overview of the Chemtrails Conspiracy Theory

The skies, often seen as the epitome of freedom and endless possibilities, have become a battleground for one of the most persistent and controversial conspiracy theories of our time—the chemtrails theory. According to proponents of this theory, the long white trails left by airplanes, traditionally known as contrails, are not mere byproducts of engine exhausts meeting cold air. Instead, these trails are allegedly a deliberate dispersal of chemical agents by governments or other powerful entities for various covert purposes.

Chemtrails theorists claim that these chemical trails contain harmful substances intended for weather modification, population control, or even mind control. The idea is that these chemicals are sprayed at high altitudes and then gradually descend to the Earth's surface, affecting the environment and human health. Despite the scientific consensus that contrails are simply water vapor and ice crystals formed by the high-altitude exhaust from jet engines, the chemtrails theory persists.

At the core of the chemtrails narrative is the belief in a shadowy cabal of elites orchestrating these operations in secret. This belief taps into broader themes of government mistrust and the fear of unseen powers manipulating the masses. The theory has found a home on various online platforms, where like-minded individuals exchange "evidence" and bolster each other's beliefs, creating an echo chamber that amplifies their message.

Understanding why the chemtrails theory remains resilient requires a deeper dive into its origins, the differences between contrails and alleged chemtrails, and the historical context that gave rise to this modern myth. Each of these elements will be explored in this chapter, setting the stage for a comprehensive examination of one of the most enduring conspiracy theories of our time.

From the initial whispers of suspicion to the full-blown movement it has become, the chemtrails theory is a fascinating case study in the power of belief, the dynamics of conspiracy thinking, and the impact of digital communication on societal fears. As we embark on this journey, it's essential to approach the subject with an open mind, recognizing the very real emotions and concerns driving those who subscribe to this theory. Whether you come away convinced or skeptical, the exploration of chemtrails is a testament to the complex interplay between science, belief, and the human need to make sense of the world around us.

Differentiating Between Contrails and Chemtrails

Contrails, or condensation trails, are a familiar sight in the skies, created by the exhaust of aircraft engines meeting the cold, moist air at high altitudes. These trails form when the hot, humid air from the jet engines mixes with the lower temperature atmosphere, causing water vapor to condense and freeze into tiny ice crystals. This process is a well-documented and understood phenomenon in atmospheric science.

Contrails can vary in appearance, sometimes dissipating quickly and other times lingering for hours, spreading into thin, cirrus-like clouds. Their behavior is influenced by a range of atmospheric conditions, including humidity, temperature, and wind patterns. This variability can often lead to misconceptions, as persistent contrails are sometimes mistaken for something more sinister by those unfamiliar with the science behind their formation.

Chemtrails, on the other hand, are purported by conspiracy theorists to be something entirely different. According to this theory, chemtrails are the result of deliberate chemical spraying from aircraft, intended for purposes such as weather modification, population control, or other covert activities. Proponents of this theory claim that chemtrails contain harmful substances like heavy metals, biological agents, or other toxins.

One of the central arguments of the chemtrails theory is the visual distinction between contrails and chemtrails. Believers assert that chemtrails have a different appearance—thicker, more persistent, and often spreading out to cover large areas of the sky. They argue that the presence of unusual patterns, crisscrossing lines, and the slow dispersion of these trails are evidence of chemical spraying.

However, scientific analysis and atmospheric studies consistently debunk these claims. The differences in contrail appearance can be attributed to varying environmental conditions rather than deliberate chemical spraying. High humidity levels can cause contrails to persist and spread, creating the crisscrossing patterns often cited by chemtrails theorists.

Moreover, extensive testing of air and water samples in areas where chemtrails are alleged to occur has repeatedly shown no evidence of unusual or harmful substances. These findings align with the understanding of contrail formation and dispersion, reinforcing the conclusion that what is seen in the skies are simply contrails, not chemtrails.

In summary, while contrails are a natural and scientifically explained phenomenon, chemtrails remain a theory unsupported by empirical evidence. The persistence of this theory highlights the importance of scientific literacy and critical thinking in distinguishing between fact and fiction. As we delve deeper into the world of chemtrails, it is crucial to keep these distinctions in mind, separating observable, natural occurrences from the realm of conspiracy.

Historical Context and Origins of the Theory

The roots of the chemtrails conspiracy theory can be traced back to the mid-1990s, a period marked by a surge in public concern over environmental and governmental activities. The theory first gained traction in the United States, where rising suspicions about government transparency and military activities were already creating fertile ground for conspiracy thinking.

One of the earliest references to chemtrails appeared in a 1996 report by the United States Air Force titled "Weather as a Force Multiplier: Owning the Weather in 2025." Although the report was a theoretical exploration of future weather modification technologies, conspiracy theorists latched onto it as supposed proof that the government was already manipulating the weather. The term "chemtrails" itself is believed to have been coined during this period, blending the words "chemical" and "trails" to describe the purported phenomenon.

As internet usage became more widespread, chemtrails theories found a new home online. Websites, forums, and social media platforms allowed like-minded individuals to share their observations and speculations, creating a sense of community among believers. The digital age facilitated the rapid spread of the theory, enabling people to connect and reinforce each other's beliefs across geographic boundaries.

A key figure in the early spread of the chemtrails theory was journalist and conspiracy theorist William Thomas. In the late 1990s, Thomas published a series of articles claiming that the United States government was conducting secretive aerosol spraying operations. His work drew significant attention and helped to solidify the narrative that chemtrails were a real and present danger. Thomas's articles were widely circulated online, and his claims were further amplified by other conspiracy theorists and alternative media outlets.

The early 2000s saw the theory gaining further momentum, with notable events like the 9/11 attacks fueling broader conspiratorial thinking. Theories about government cover-ups and hidden agendas became more pervasive, and chemtrails fit neatly into this expanding web of distrust. Books, documentaries, and radio shows dedicated to exploring the chemtrails phenomenon began to appear, each adding new layers of speculation and alleged evidence to the growing narrative.

Despite numerous scientific debunkings, the chemtrails theory continued to attract followers. Skeptics and experts pointed out that the atmospheric conditions required for contrail formation were well understood and that there was no credible evidence to support claims of chemical spraying. Nonetheless, the theory persisted, driven by a combination of anecdotal evidence, selective interpretations of scientific studies, and a general mistrust of official explanations.

Understanding the historical context and origins of the chemtrails theory is crucial for grasping why it has endured for so long. It is a testament to the power of narrative and the influence of collective belief, demonstrating how conspiracy theories can take root and flourish in an environment of uncertainty and mistrust. As we continue to explore the world of chemtrails, this historical perspective provides a foundation for understanding the complex interplay between fact, fiction, and the human desire to uncover hidden truths.

Spread of the Theory and Key Proponents

As the chemtrails theory emerged from its infancy in the 1990s, it quickly gained momentum, fueled by a growing distrust in government and increased access to global communication through the internet. The theory spread rapidly, with early proponents harnessing the power of online platforms to amplify their message and connect with like-minded individuals across the world.

Websites dedicated to uncovering "the truth" about chemtrails began to pop up, each offering a space for believers to share their observations, concerns, and supposed evidence. Discussion forums and email chains circulated anecdotal reports and speculative articles, creating an ever-growing archive of chemtrails-related content. This digital age of information exchange allowed the theory to flourish, reaching audiences far beyond its initial geographic confines.

Key figures emerged as leaders in the chemtrails movement, each contributing to the theory's credibility and expanding its reach. Among them was Clifford Carnicom, a former government scientist who dedicated his research to uncovering the truth behind chemtrails. Carnicom's detailed reports and scientific-sounding language lent an air of legitimacy to the theory, attracting followers who sought validation from authoritative voices.

Another prominent advocate was journalist and author Jim Marrs, known for his work on various conspiracy theories. Marrs published extensively on chemtrails, drawing connections between weather modification programs, government secrecy, and the alleged harmful effects of chemtrails on human health and the environment. His books and public appearances further popularized the theory, bringing it into the mainstream consciousness.

Social media platforms like Facebook, YouTube, and Twitter became vital tools for spreading the chemtrails narrative. Videos showing "evidence" of chemtrails crisscrossing the sky, testimonials from supposed whistleblowers, and dramatic music underscoring ominous warnings about government cover-ups captivated audiences. The visual nature of these platforms allowed for a more compelling and immediate presentation of the theory, often bypassing critical examination in favor of emotional appeal.

The role of alternative media cannot be underestimated in the spread of the chemtrails theory. Independent radio shows, podcasts, and online publications dedicated significant airtime to discussing chemtrails, often framing the theory within a broader context of distrust towards official narratives. This alternative media ecosystem created an echo chamber where the theory could thrive, shielded from mainstream debunking efforts and scientific scrutiny.

By the mid-2000s, the chemtrails theory had become a global phenomenon. It found receptive audiences in various countries, each adapting the narrative to fit local contexts and concerns. In some regions, chemtrails were linked to weather anomalies; in others, they were blamed for health crises. This adaptability allowed the theory to maintain its relevance and continue its spread despite mounting scientific evidence to the contrary.

The proliferation of the chemtrails theory demonstrates the powerful influence of community and shared belief in sustaining conspiracy theories. As individuals found solidarity and validation within the chemtrails movement, the theory's grip on the collective imagination strengthened. Understanding this process is key to comprehending why such theories persist and how they evolve in the digital age.

Public Perception and Initial Skepticism

When the chemtrails theory first emerged, it was met with a mix of curiosity, fear, and skepticism. Public perception was deeply divided, with some individuals quickly adopting the theory as an alarming truth, while others dismissed it outright as unfounded paranoia. This polarized reaction was indicative of the broader landscape of conspiracy theories and the varying levels of public trust in government and scientific institutions.

For those who believed in the chemtrails theory, the idea tapped into existing anxieties about environmental degradation, governmental overreach, and the potential for unseen threats. The imagery of skies crisscrossed with chemical trails evoked a sense of vulnerability, suggesting that everyday activities, such as air travel, were being used to perpetrate large-scale deception. This fear was amplified by the lack of transparency and the occasional missteps of governmental agencies, which sometimes fueled suspicion rather than alleviating it.

On the other hand, initial skepticism came from both the scientific community and members of the general public who trusted established scientific explanations. Meteorologists, atmospheric scientists, and aviation experts were quick to point out the flaws in the chemtrails theory, emphasizing the well-understood processes behind contrail formation. These experts highlighted the lack of credible evidence to support claims of chemical spraying and stressed the importance of relying on peer-reviewed research rather than anecdotal reports and speculative assertions.

Early debunking efforts included educational campaigns aimed at explaining the science of contrails to the public. Scientists and skeptics used platforms like news articles, television segments, and public lectures to disseminate accurate information about atmospheric phenomena. These efforts sought to counteract the spread of misinformation and reassure the public that the trails seen in the sky were a natural byproduct of jet engine exhaust, not a covert chemical operation.

Despite these efforts, the chemtrails theory continued to gain traction, partly due to the emotional and visually compelling nature of the claims. Photographs and videos showing persistent trails in the sky were shared widely on social media, often accompanied by dramatic narratives and calls to action. This visual evidence seemed persuasive to many, even in the face of scientific rebuttals, highlighting the challenge of combating misinformation in the digital age.

Public perception of chemtrails has evolved over time, with waves of belief and skepticism ebbing and flowing in response to new claims, debunkings, and societal events. The theory has become part of a broader tapestry of modern conspiracy beliefs, reflecting ongoing tensions between institutional authority and grassroots questioning. As we move forward in this book, understanding these initial reactions and the reasons behind them will be crucial in exploring the enduring appeal of the chemtrails theory and its impact on public discourse.

Chapter 2: The Science of Contrails

Formation of Contrails

Contrails, short for condensation trails, are streaks of condensed water vapor created in the atmosphere by the exhaust of aircraft engines. The formation of contrails is a result of a fascinating interplay between the physics of high-altitude flight and the unique conditions of the upper atmosphere. To understand how contrails are formed, one must first grasp the fundamental principles of jet propulsion and atmospheric dynamics.

When an airplane travels at high altitudes, typically above 26,000 feet, its engines expel hot exhaust gases. These gases contain a significant amount of water vapor, a byproduct of the combustion process. At these altitudes, the ambient air is extremely cold, often reaching temperatures as low as -40 degrees Celsius or lower. As the hot, moist exhaust gases exit the engine and encounter the frigid ambient air, the water vapor in the exhaust cools rapidly and condenses into tiny water droplets or ice crystals.

This rapid condensation occurs because the cold air cannot hold as much moisture as the warm exhaust gases. The sudden drop in temperature causes the water vapor to transition from a gaseous state to a liquid or solid state almost instantaneously. The result is the formation of a visible trail of condensed water vapor or ice crystals behind the aircraft, commonly known as a contrail.

Contrails can vary in appearance based on several factors, including the temperature and humidity of the surrounding air. In extremely cold and dry conditions, contrails may form as thin, short-lived streaks that dissipate quickly. These fleeting contrails are often referred to as "short-lived contrails" and are typically seen trailing behind an aircraft for only a few seconds or minutes before they vanish.

In contrast, when the atmospheric conditions are more humid, the water droplets or ice crystals in the contrail can persist for a longer period. These persistent contrails may linger in the sky for several hours, gradually spreading out and forming extensive, wispy clouds known as cirrus clouds. The persistence and spread of contrails are influenced by the level of humidity at the altitude where the aircraft is flying. Higher humidity levels allow the contrails to remain visible for a longer time and can lead to the formation of larger cloud formations.

The formation of contrails is also influenced by the type of aircraft and its operational parameters. Modern jet engines, which are highly efficient and operate at higher temperatures, are more likely to produce contrails compared to older engines. Additionally, the altitude and speed at which an aircraft flies can affect the characteristics of the contrails it produces.

Understanding the science behind contrail formation is essential for distinguishing between natural atmospheric phenomena and unfounded conspiracy theories. Contrails are a well-documented consequence of high-altitude flight and are governed by the laws of physics and atmospheric science. By recognizing the factors that contribute to the formation and persistence of contrails, we can better appreciate the complex interplay between human activity and the natural environment. This knowledge also serves as a foundation for debunking misconceptions and clarifying the scientific realities behind the visible trails left in the sky by passing aircraft.

Atmospheric Conditions and Contrail Persistence

Contrails, while primarily a product of jet exhaust, owe their varying appearances and durations to the intricate dance of atmospheric conditions at high altitudes. To fully appreciate why some contrails vanish almost instantaneously while others linger and spread, it's essential to delve into the specific environmental factors at play.

The persistence of contrails is largely dictated by the ambient temperature and humidity levels at the altitude where they form. In the upper troposphere, where commercial jets typically cruise, temperatures can plummet to -40 degrees Celsius or even lower. This extreme cold is a critical factor in the formation of contrails. When hot, moist exhaust gases from an aircraft engine are expelled into this frigid air, they cool rapidly, causing the water vapor to condense into liquid droplets or, more commonly, freeze into ice crystals.

Humidity is the next crucial factor. The upper atmosphere can vary significantly in its moisture content. In conditions where the air is very dry, the ice crystals in a contrail will quickly sublimate, turning from solid directly back to gas. This results in short-lived contrails that may only last a few seconds to minutes. On the other hand, when the relative humidity is high, these ice crystals can persist. They absorb more moisture from the surrounding air, causing the contrail to expand and spread, sometimes merging with other contrails and forming extensive cirrus cloud-like formations.

Pressure also plays a role. At higher altitudes, the pressure is much lower than at the Earth's surface. This lower pressure means that the air can hold less water vapor. As a result, the process of condensation and freezing happens more readily, contributing to the initial formation of contrails. However, for these contrails to persist, the ambient conditions must allow the ice crystals to remain stable without sublimating too quickly.

Scientific studies have provided ample data to support these mechanisms. For instance, research utilizing satellite observations and atmospheric modeling has shown that persistent contrails are more likely to form in regions with higher relative humidity and colder temperatures. These studies have also helped to map out the typical altitudes and geographic locations where contrails are most frequently observed.

Meteorologists and atmospheric scientists use tools like radiosondes and weather balloons to measure temperature, humidity, and pressure at various altitudes. These measurements are crucial for predicting contrail formation and persistence. For example, if a weather forecast predicts a moist and cold upper atmosphere, it's likely that persistent contrails will form. Conversely, if the forecast indicates dry conditions, contrails will likely be short-lived.

The interaction of these atmospheric conditions not only determines the lifespan of a contrail but also its appearance. Persistent contrails can undergo a process called spreading, where wind shear—variations in wind speed and direction at different altitudes—stretches and diffuses the contrail, causing it to expand and sometimes cover large portions of the sky. This phenomenon is often mistaken for deliberate chemical spraying by those unaware of the underlying science.

Understanding the atmospheric conditions that influence contrail persistence dispels much of the mystery and misinformation surrounding these sky-high streaks. It underscores the importance of scientific literacy in interpreting natural phenomena and challenges the unfounded claims of chemtrails by presenting clear, evidence-based explanations. This knowledge also highlights the delicate balance of conditions required for contrails to form and persist, offering a glimpse into the complex dynamics of our planet's atmosphere.

Visual Characteristics of Contrails

The intricate patterns of contrails that crisscross the sky are not only a testament to the marvels of modern aviation but also an intriguing subject of atmospheric science. Understanding why contrails look the way they do requires a closer look at the variables that influence their visual characteristics, including the size, shape, and spread of the ice crystals that compose them.

At their core, contrails are composed of tiny ice crystals formed from the water vapor expelled by aircraft engines. The initial appearance of these contrails is determined by the temperature and humidity at the altitude where the plane is flying. In extremely cold and relatively moist conditions, contrails form almost immediately behind the aircraft, appearing as thin, sharp lines against the backdrop of the sky. These trails can quickly expand if the surrounding air is sufficiently humid, leading to broader, more diffuse patterns.

The spread of contrails is heavily influenced by wind patterns at high altitudes. When an aircraft moves through an atmosphere with strong wind shear—differences in wind speed and direction at various altitudes—the contrails can be stretched and twisted into a variety of shapes. These wind-induced distortions can create the crisscrossing or spiraling patterns often observed. As the ice crystals drift and disperse, they can merge with other contrails, forming extensive sheets of cirrus clouds that can cover large areas.

The duration and visibility of contrails are also affected by the size of the ice crystals themselves. In colder, less humid conditions, the ice crystals tend to be smaller and more prone to sublimation, where they transition directly from a solid state back to vapor. This results in shorter-lived contrails that dissipate rapidly. Conversely, in more humid environments, the ice crystals can grow larger and persist for extended periods, creating contrails that can last for hours.

Additionally, contrails can exhibit a range of optical phenomena due to the interaction of light with the ice crystals. This includes the bright, white appearance of fresh contrails, caused by the scattering of sunlight by the ice particles. At certain angles, contrails can also produce halos and iridescent colors, similar to those seen in other types of ice crystal clouds. These optical effects can be particularly striking during sunrise or sunset when the angle of the sunlight is low, casting dramatic hues across the sky.

It's important to note that contrails are not unique to modern jet aircraft. Similar trails have been observed behind high-altitude piston-engine aircraft and even rockets, as long as the necessary atmospheric conditions are met. This further underscores that contrails are a natural byproduct of high-altitude flight, not evidence of any deliberate chemical spraying.

Understanding the visual characteristics of contrails helps demystify the various shapes and behaviors they exhibit. It highlights the complex interplay of environmental factors that shape these ephemeral sky patterns and dispels misconceptions about their origins. By recognizing the science behind contrail formation and appearance, we can appreciate the beauty and intricacy of these fleeting creations while grounding our understanding in factual, evidence-based knowledge.

Impact on Climate and Weather

Contrails, while a fascinating subject of scientific study on their own, also have broader implications for our climate and weather patterns. These icy trails left by high-flying aircraft contribute to atmospheric processes in ways that are still being explored by scientists, and their impact on both local and global scales is significant.

One of the key areas of study is the effect of contrails on Earth's radiation balance. Contrails, particularly those that persist and spread into cirrus-like clouds, can influence the amount of solar radiation that reaches the Earth's surface. During the day, these contrail-induced clouds reflect some of the incoming sunlight back into space, a process known as "albedo" effect, which has a cooling effect on the atmosphere. However, at night, these same contrails trap outgoing infrared radiation, acting as a blanket that retains heat and contributes to warming. This dual effect means that the overall impact of contrails on climate can be complex, with both cooling and warming influences.

The extent of this impact was dramatically highlighted during the grounding of commercial flights in the United States following the September 11, 2001, terrorist attacks. During the three-day period when airplanes were largely absent from the skies, scientists observed noticeable changes in the diurnal temperature range (the difference between daytime and nighttime temperatures). The absence of contrails allowed more solar radiation to reach the surface during the day and more infrared radiation to escape at night, leading to greater temperature fluctuations. This provided a rare opportunity to study the impact of aviation on the atmosphere and underscored the significant role that contrails play in climate dynamics.

Further research into the climatic impact of contrails focuses on their contribution to global dimming and global warming. Global dimming refers to the reduction in the amount of sunlight reaching the Earth's surface, partly due to contrails and other human-made aerosols. This phenomenon can affect weather patterns, agricultural productivity, and even the hydrological cycle. By reflecting sunlight, contrails can contribute to cooler surface temperatures in some regions, but their overall contribution to global warming, through their heat-trapping effects, is a subject of ongoing study.

In addition to their climatic effects, contrails can influence local weather patterns. Persistent contrails that spread into extensive cloud layers can modify cloud cover, impacting precipitation and cloud dynamics. Some studies suggest that regions with heavy air traffic may experience altered weather patterns due to the increased presence of contrails. These changes can affect everything from local temperatures to the distribution of rainfall, although the exact mechanisms and outcomes are still being investigated.

The scientific consensus on contrails' environmental impact is evolving as new data and methodologies become available. Advanced satellite monitoring, atmospheric modeling, and field experiments are helping to refine our understanding of how contrails interact with the climate system. This research is crucial for developing more sustainable aviation practices and mitigating the environmental footprint of air travel.

In conclusion, while contrails might appear as mere streaks in the sky, their impact on climate and weather is far-reaching. By influencing radiation balance, contributing to global dimming, and modifying local weather patterns, contrails play a significant role in the atmospheric processes that shape our environment. Continued research in this field is essential for balancing the benefits of aviation with the need to protect our planet's climate and ecosystems.

Distinguishing Contrails from Chemtrails

The distinction between contrails and the so-called chemtrails is crucial to understanding why the latter remains a controversial topic, despite overwhelming scientific evidence to the contrary. While contrails are a well-understood phenomenon resulting from the physics of high-altitude flight, chemtrails are a product of conspiracy theories that misinterpret these natural occurrences.

Contrails, as established, are formed when water vapor from an aircraft's exhaust condenses into ice crystals in the cold upper atmosphere. These trails can appear as thin lines that either dissipate quickly or persist and spread, depending on atmospheric conditions. The science behind contrails is supported by decades of research and observation, with clear explanations grounded in the principles of meteorology and atmospheric science.

Chemtrails, on the other hand, are purported to be deliberate chemical releases by aircraft for purposes such as weather manipulation, population control, or other nefarious activities. Proponents of the chemtrails theory point to the appearance and persistence of certain contrails as evidence, claiming they contain harmful chemicals. However, these claims lack credible scientific backing and often rely on anecdotal observations and misinterpretations of scientific data.

To distinguish contrails from the alleged chemtrails, one must rely on several key scientific criteria:

Formation Process: Contrails form through the condensation of water vapor in aircraft exhaust under specific atmospheric conditions. Chemtrail proponents often ignore this well-documented process, instead attributing the formation of persistent contrails to chemical spraying. Understanding the basic science behind contrail formation debunks many of the claims made by chemtrail theorists.

Chemical Composition: Extensive testing of air and water samples in areas purportedly affected by chemtrails has consistently failed to detect unusual or harmful substances. Independent studies and governmental agencies have conducted numerous analyses, finding no evidence to support the presence of chemicals beyond what is naturally present in the environment.

Atmospheric Conditions: The persistence and spread of contrails are influenced by temperature, humidity, and wind patterns at high altitudes. These conditions can cause contrails to linger and spread, creating patterns that chemtrail proponents often cite as evidence of chemical spraying. Recognizing the role of these atmospheric variables is essential for understanding why some contrails persist longer than others.

Scientific Consensus: The scientific community overwhelmingly supports the explanation of contrails as a natural byproduct of aviation. Peer-reviewed studies, atmospheric models, and expert opinions consistently debunk the chemtrails theory. Leading meteorological and environmental organizations affirm that contrails pose no significant health or environmental risks.

Misinformation and Misinterpretation: Many chemtrail claims stem from a misunderstanding or misinterpretation of scientific data. Photographs and videos showing persistent contrails are often used as "evidence," but without the context of atmospheric science, these images can be misleading. Educating the public about the true nature of contrails is crucial for dispelling myths and promoting scientific literacy.

In conclusion, the distinction between contrails and chemtrails is rooted in the application of rigorous scientific principles versus the propagation of unsubstantiated conspiracy theories. By understanding the formation, composition, and behavior of contrails, we can confidently dismiss the chemtrails narrative as a misunderstanding of natural atmospheric processes. Promoting accurate information and critical thinking is key to addressing the misconceptions that fuel the chemtrails theory and ensuring that our understanding of the skies above remains grounded in science.

Chapter 3: The Birth of a Conspiracy

Early Proponents and Spread of the Chemtrails Theory
The genesis of the chemtrails conspiracy theory can be traced back to the mid-1990s, an era marked by the proliferation of the internet and growing public suspicion towards government activities. This period saw a fertile ground for the germination of various conspiracy theories, with chemtrails emerging as one of the most enduring.

One of the earliest and most influential proponents of the chemtrails theory was journalist William Thomas. In 1997, Thomas published a series of articles and gave several interviews asserting that the trails left by jet aircraft contained harmful chemicals. His claims were based on anecdotal evidence and misinterpretations of atmospheric phenomena, but they resonated with a segment of the public already inclined to distrust official explanations.

Thomas's assertions were soon picked up by other conspiracy theorists, creating a snowball effect. Websites dedicated to exposing the "truth" about chemtrails began to appear, and internet forums buzzed with discussions about mysterious illnesses, unusual weather patterns, and the alleged spraying activities of governments and shadowy organizations. These early digital platforms played a crucial role in spreading the chemtrails theory, allowing isolated anecdotes and speculative interpretations to reach a global audience.

In addition to Thomas, another significant figure in the early days of the chemtrails theory was Clifford Carnicom. A former government scientist, Carnicom devoted much of his time to researching and promoting the idea that chemtrails were a deliberate, large-scale operation to manipulate the environment and human health. Carnicom's technical background lent an air of credibility to his claims, and his detailed reports and scientific-sounding language helped to solidify the narrative.

As the theory spread, it began to take on a life of its own, incorporating elements from other conspiracy theories. Claims of weather modification, mind control, and even population reduction became intertwined with the chemtrails narrative. The theory's flexibility allowed it to adapt and evolve, incorporating new "evidence" and adjusting to debunking efforts. This adaptability made chemtrails particularly resilient in the face of scientific rebuttals and official denials.

The internet's role in the spread of the chemtrails theory cannot be overstated. Platforms like YouTube, Facebook, and Twitter allowed proponents to share videos, photos, and personal testimonies, creating an immersive and persuasive multimedia experience. These platforms also enabled the rapid dissemination of misinformation, with viral posts often outpacing efforts to correct or debunk them. The visual nature of the content shared—images of crisscrossed skies, lingering trails, and purported evidence of chemical residues—added a powerful emotional component that appealed to viewers' fears and suspicions.

The early spread of the chemtrails theory exemplifies how modern technology can amplify and perpetuate misinformation. The combination of charismatic proponents, compelling visuals, and a receptive audience created a perfect storm for the theory's growth. Understanding these early dynamics is crucial for appreciating why the chemtrails theory has persisted and continues to captivate a segment of the public. As we delve deeper into the world of chemtrails, this historical perspective will help us unravel the complex interplay between belief, media, and the human propensity for conspiracy thinking.

Key Events and Publications That Fueled the Theory

The chemtrails conspiracy theory didn't gain widespread attention overnight; rather, it was the result of a series of key events and publications that slowly but steadily built its momentum. Understanding these milestones provides insight into how the theory evolved from a fringe idea into a widely discussed topic.

One of the pivotal moments in the spread of the chemtrails theory occurred in 1996 when the United States Air Force published a report titled "Weather as a Force Multiplier: Owning the Weather in 2025." Although the report was speculative and intended to explore future technological possibilities, conspiracy theorists seized upon its contents. They interpreted the document as evidence that the government was already engaging in large-scale weather modification projects. This misinterpretation fueled the belief that the trails left by airplanes were part of a covert operation to manipulate the weather.

Following the publication of this report, several prominent figures in the conspiracy community began to draw connections between it and the appearance of persistent contrails. Radio host Art Bell, known for his show "Coast to Coast AM," which delved into various paranormal and conspiracy topics, frequently discussed chemtrails. Bell's platform reached millions of listeners, amplifying the theory and giving it a broader audience. His interviews with guests who claimed insider knowledge about chemtrails further cemented the idea in the public's mind.

Another significant contribution came from author and conspiracy theorist Jim Marrs. In his book "Rule by Secrecy," published in 2000, Marrs dedicated a chapter to chemtrails, linking them to a broader narrative of government control and hidden agendas. Marrs' work, which appealed to readers already skeptical of official narratives, provided a comprehensive framework that intertwined chemtrails with other conspiracy theories. This holistic approach made the chemtrails theory more compelling, as it connected various dots that seemed unrelated at first glance.

Television also played a crucial role in propagating the chemtrails theory. A 2005 episode of the popular investigative series "MysteryQuest" titled "Chemtrails" brought the theory into the mainstream. The episode presented interviews with alleged experts and eyewitnesses who claimed to have evidence of chemical spraying. Although the show was designed to entertain rather than inform, its dramatized portrayal of the chemtrails phenomenon lent an air of legitimacy to the theory, drawing in viewers who might not have encountered it otherwise.

In the print media, newspapers and magazines occasionally featured articles on chemtrails, often in a sensationalist manner. These articles typically highlighted personal testimonies and unexplained phenomena, without delving deeply into the scientific explanations for contrails. The lack of rigorous investigative journalism on the topic allowed misconceptions to flourish, as readers were left with more questions than answers.

The rise of social media in the mid-2000s provided a new and powerful platform for the spread of the chemtrails theory. Sites like YouTube became repositories for documentaries, personal testimonies, and amateur investigations into chemtrails. These videos, often featuring dramatic music and visually striking footage of the sky, attracted millions of views. The comment sections became hubs for believers to share their experiences and reinforce each other's beliefs, creating a sense of community and shared purpose.

Each of these key events and publications contributed to the chemtrails theory's growth, intertwining it with a broader narrative of distrust towards government and authority. By examining these milestones, we gain a clearer understanding of how the theory was constructed and why it continues to resonate with certain segments of the population. As we delve further into the world of chemtrails, these historical contexts will provide a foundation for analyzing the theory's persistence and evolution.

Role of the Internet and Social Media in Spreading the Theory

In the late 1990s and early 2000s, the burgeoning internet and the rise of social media revolutionized how information was shared and consumed, playing a pivotal role in amplifying the chemtrails conspiracy theory. The internet provided an unprecedented platform for individuals to disseminate ideas, connect with like-minded individuals, and challenge mainstream narratives—all of which were crucial to the spread of the chemtrails theory.

One of the most significant ways the internet facilitated the spread of chemtrails was through the creation of dedicated websites and forums. These digital spaces became hubs for the exchange of information, where believers could share their observations, theories, and supposed evidence without the filters of mainstream media. Websites like Carnicom Institute and Chemtrail Central became go-to sources for those seeking to validate their beliefs or find a community of fellow theorists. These platforms hosted a plethora of articles, photographs, videos, and personal testimonies, creating an extensive archive that lent an air of legitimacy to the theory.

Forums and message boards further fostered a sense of community among chemtrails believers. On platforms like Above Top Secret and Godlike Productions, users could engage in detailed discussions about their sightings, exchange tips on identifying chemtrails, and share their interpretations of scientific studies and government documents. These interactions reinforced the sense of camaraderie and collective purpose, making it easier for individuals to reject debunking efforts and maintain their belief in the theory.

The advent of social media exponentially increased the reach and impact of the chemtrails theory. Platforms like Facebook, YouTube, and Twitter allowed users to share content with a global audience instantly. Videos purporting to show chemtrails in action, complete with dramatic commentary and music, went viral, attracting millions of views. These videos often featured interviews with self-proclaimed experts and whistleblowers, adding a veneer of credibility to the claims. The visual and emotional appeal of these videos made them highly effective at capturing viewers' attention and persuading them of the theory's validity.

Facebook groups dedicated to chemtrails also played a crucial role in spreading the theory. These groups, some with thousands of members, functioned as echo chambers where individuals could reinforce each other's beliefs and dismiss contradictory evidence. Administrators of these groups curated content to support the narrative, and members often shared personal stories of health issues or environmental changes they attributed to chemtrails. This constant reinforcement helped to solidify the belief that chemtrails were a real and present danger.

Twitter hashtags like #chemtrails and #geoengineering enabled users to join a broader conversation and connect with others who shared their concerns. These hashtags facilitated the rapid spread of information and disinformation, allowing chemtrails content to reach a wide audience quickly. The succinct nature of tweets made it easy to share striking images and provocative statements, further embedding the theory in the public consciousness.

The role of the internet and social media in spreading the chemtrails theory highlights the powerful influence of digital platforms on modern conspiracy theories. By providing spaces for like-minded individuals to gather, share information, and reinforce each other's beliefs, these platforms have helped to sustain and amplify the chemtrails narrative. Understanding this dynamic is essential for comprehending how and why the theory has persisted despite overwhelming scientific evidence to the contrary. As we continue to explore the world of chemtrails, the impact of digital communication will remain a central theme in unraveling the complex interplay between belief, media, and the search for truth.

Early Opposition and Media Coverage

From its inception, the chemtrails conspiracy theory has faced significant opposition from scientists, government officials, and skeptics who have consistently sought to debunk its claims. This section explores the early efforts to counter the theory and the role of media coverage in shaping public perception.

As the chemtrails theory began to gain traction in the late 1990s, the scientific community quickly responded with efforts to educate the public about the nature of contrails and the lack of evidence supporting the chemtrails claims. Meteorologists and atmospheric scientists emphasized that contrails are a well-understood phenomenon resulting from the condensation of water vapor in aircraft exhaust at high altitudes. They pointed out that the conditions required for contrail formation—cold temperatures and high humidity—were well documented and explained why some contrails persisted longer than others.

One of the earliest and most comprehensive scientific rebuttals came from the United States Air Force. In response to growing public inquiries and concerns, the Air Force published a fact sheet in 2000 titled "Chemtrails—Fact or Fiction?" The document clearly outlined the scientific basis for contrail formation and addressed common misconceptions perpetuated by the chemtrails theory. The fact sheet also included statements from various scientific organizations, all of which confirmed that there was no evidence of chemical spraying by aircraft.

Despite these efforts, the chemtrails theory continued to thrive, partly due to sensationalist media coverage. Mainstream media outlets, recognizing the public's fascination with conspiracy theories, often featured stories about chemtrails in a way that balanced skepticism with intrigue. Headlines such as "Are We Being Sprayed?" and "The Secret War in the Skies" captured readers' attention and fueled further speculation. These articles often included interviews with both proponents and opponents of the theory, giving equal weight to both sides and creating a false equivalence that undermined the scientific consensus.

Television programs also played a role in perpetuating the chemtrails narrative. Shows like "MysteryQuest" and "Conspiracy Theory with Jesse Ventura" dedicated episodes to exploring chemtrails, presenting them as mysterious phenomena worthy of investigation. While these programs were intended for entertainment, their dramatized portrayal of the theory lent it an air of legitimacy, encouraging viewers to question the official explanations provided by scientists and government agencies.

Alternative media outlets, such as blogs and independent news websites, were even more instrumental in spreading the chemtrails theory. These platforms often positioned themselves as truth-seekers, challenging mainstream narratives and providing a space for alternative viewpoints. Articles and videos on these sites frequently framed the chemtrails theory within a broader context of government mistrust and secretive agendas, resonating with readers who were already skeptical of officialdom.

The impact of early opposition and media coverage on the chemtrails theory highlights the challenges of combating misinformation in the digital age. While scientists and skeptics worked diligently to debunk the theory, their efforts were often overshadowed by the allure of sensationalist stories and the emotional appeal of conspiracy thinking. As we continue to examine the evolution of the chemtrails theory, it is essential to understand the complex interplay between scientific evidence, media representation, and public perception. This dynamic has shaped not only the trajectory of the chemtrails narrative but also the broader landscape of modern conspiracy theories.

The Role of Whistleblowers and Insiders

The persistence and proliferation of the chemtrails conspiracy theory owe much to the accounts of self-proclaimed whistleblowers and insiders who have claimed to have firsthand knowledge of covert chemical spraying operations. These individuals often present themselves as former government or military personnel, scientists, or aviation experts who have broken ranks to expose what they describe as a massive cover-up. Their testimonies have added a veneer of credibility to the theory, further entrenching it in the minds of believers.

One of the most influential figures in this regard is a man known by the pseudonym "Dr. Bill Deagle." Deagle, a former medical doctor, emerged as a prominent voice in the chemtrails community in the early 2000s. He claimed to have insider knowledge of various government programs, including weather modification and chemical spraying operations. Deagle's dramatic presentations and detailed allegations captivated audiences at conferences and in online videos, making him a key figure in the chemtrails narrative. However, his lack of verifiable credentials and the outlandish nature of some of his claims have led many to question his credibility.

Another notable figure is Kristen Meghan, a former industrial hygienist and environmental specialist for the U.S. Air Force. Meghan has publicly stated that during her tenure with the Air Force, she discovered evidence of large quantities of chemicals being purchased and used in operations that she believed were related to chemtrails. Her professional background and articulate presentations have made her a compelling spokesperson for the theory, although her claims have not been substantiated by independent investigations.

The accounts of these and other alleged whistleblowers are often featured prominently in documentaries, online videos, and alternative media reports. These testimonies typically share several common elements: they describe secretive programs, emphasize the whistleblower's personal risk in coming forward, and present anecdotal evidence rather than concrete proof. Despite the lack of verifiable evidence, these stories resonate with believers because they align with preexisting suspicions about government secrecy and malfeasance.

In addition to individual whistleblowers, various purported leaks of government documents have been cited as evidence of chemtrails. For instance, proponents often reference documents from Project Popeye, a weather modification program during the Vietnam War, as proof that the U.S. government has engaged in environmental manipulation. While Project Popeye and other historical weather modification efforts are well-documented, there is no evidence linking these programs to the large-scale chemical spraying described by chemtrails theorists.

The influence of whistleblowers and insiders on the chemtrails theory underscores the power of personal testimony and anecdotal evidence in shaping public perception. In the absence of concrete proof, the stories of individuals claiming insider knowledge can be persuasive, especially when they play into broader narratives of distrust and skepticism toward authority. These accounts add a human element to the theory, making it more relatable and emotionally impactful for those who are already inclined to believe.

However, the reliance on unverifiable testimonies and anecdotal evidence highlights a significant weakness in the chemtrails theory. Without concrete, independently verifiable proof, the theory remains speculative and unsupported by the broader scientific community. As we continue to explore the world of chemtrails, it is essential to critically evaluate the sources of information and consider the motivations and credibility of those who claim to have insider knowledge. This critical approach is key to separating fact from fiction and understanding the complex dynamics that sustain the chemtrails narrative.

Chapter 4: Alleged Purposes of Chemtrails

Weather Modification Claims

One of the most prominent and widely discussed purposes attributed to chemtrails is weather modification. Proponents of this theory assert that the trails left by aircraft are not simply contrails—composed of water vapor and ice crystals—but are instead loaded with chemicals designed to influence weather patterns. This belief taps into a long history of human attempts to control and modify the weather, giving it a veneer of plausibility that resonates with the public.

The idea of weather modification is not without historical precedent. Since the mid-20th century, various governments and research organizations have experimented with techniques to influence weather conditions. One of the earliest and most well-known methods is cloud seeding, which involves dispersing substances such as silver iodide or sodium chloride into the atmosphere to encourage the formation of clouds and precipitation. Cloud seeding has been used to alleviate drought conditions, enhance snowfall in ski resorts, and reduce the impact of hailstorms on crops.

These historical examples provide a foundation for the claims made by chemtrails theorists. They argue that modern-day chemtrails are an extension of these earlier weather modification efforts, scaled up and made more sophisticated through advances in technology and atmospheric science. The substances purportedly sprayed from aircraft, according to proponents, include not only silver iodide but also a variety of other chemicals designed to manipulate the weather in more precise and impactful ways.

Believers in the weather modification theory point to several pieces of "evidence" to support their claims. They highlight patents for weather modification technologies, which indeed exist but are often misinterpreted or taken out of context. These patents, covering a range of techniques and substances, are cited as proof that the government or other powerful entities are actively pursuing large-scale weather modification programs. Additionally, unusual weather patterns, such as prolonged droughts or intense storms, are often attributed to chemtrails, with theorists arguing that these events are not natural but rather the result of deliberate intervention.

Moreover, personal testimonies and anecdotal observations play a significant role in bolstering the weather modification claims. Individuals report seeing unusual cloud formations, persistent trails that linger and spread, and changes in weather conditions that they believe are linked to chemtrails. These observations are shared widely on social media platforms and alternative media sites, creating a feedback loop that reinforces the belief in chemtrails and their supposed purposes.

However, the scientific community remains highly skeptical of these claims. While weather modification research and technology exist, there is no credible evidence to support the idea that chemtrails are part of a covert weather modification program. The atmospheric conditions required for contrail formation and persistence are well understood, and the variability in their appearance is attributed to natural factors such as temperature, humidity, and wind patterns at high altitudes.

Furthermore, extensive testing and analysis of air and water samples in areas where chemtrails are alleged to occur have found no evidence of unusual or harmful substances. These findings align with the understanding that contrails are composed of water vapor and ice crystals, not chemicals designed to modify the weather.

In summary, while the idea of weather modification through chemtrails taps into a historical context of genuine scientific experimentation, it remains unsupported by empirical evidence. The persistence of this theory highlights the challenge of distinguishing between legitimate scientific inquiry and unfounded conspiracy claims. As we explore other alleged purposes of chemtrails, the importance of critical thinking and reliance on credible sources becomes increasingly evident.

Population Control Theories

Another compelling aspect of the chemtrails conspiracy theory is the belief that chemtrails are being used for population control or depopulation. Proponents of this theory argue that the chemicals purportedly dispersed in the atmosphere are intended to have deleterious effects on human health, ultimately leading to reduced population numbers. This idea leverages longstanding fears about government and elite manipulation of the masses, tapping into deep-seated anxieties about autonomy and personal safety.

Central to the population control theory are claims about the substances allegedly being sprayed from aircraft. According to theorists, these chemicals include heavy metals like aluminum and barium, as well as biological agents and other toxins. Believers assert that these substances can cause a wide range of health problems, including respiratory issues, neurological disorders, and weakened immune systems. They argue that the cumulative effect of exposure to these chemicals is a gradual deterioration of public health, contributing to higher mortality rates and lower birth rates.

Proponents of the population control theory often point to various motivations that they attribute to governments and elites. These motivations range from controlling overpopulation and managing limited resources to reducing healthcare costs and maintaining political power. The theory posits that by controlling population growth, those in power can ensure a more manageable and subservient populace, thereby safeguarding their own interests.

The evidence cited by population control theorists is often anecdotal and circumstantial. Many point to increases in certain health conditions, such as asthma and Alzheimer's disease, as evidence of the harmful effects of chemtrails. They also cite personal testimonies of individuals who claim to have experienced health problems that they attribute to chemtrail exposure. These accounts are frequently shared on social media and alternative media platforms, creating a sense of urgency and validation among believers.

However, the scientific community has consistently debunked these claims. Extensive research and testing have shown no link between contrails and adverse health effects. Studies conducted by environmental and public health agencies have found that the concentrations of heavy metals and other substances in the environment are well within safe limits and are not associated with aircraft emissions. Additionally, the purported symptoms and health conditions cited by chemtrails theorists can be explained by other environmental and lifestyle factors.

One of the most significant challenges to the population control theory is the sheer scale and complexity of such an operation. Implementing a coordinated, global program of chemical spraying would require immense logistical capabilities, extensive secrecy, and the cooperation of numerous governments and agencies—an improbable scenario given the diversity of political systems and interests worldwide. Moreover, the widespread dissemination of chemicals in the atmosphere would not discriminate among populations, making it unlikely to serve the targeted, controlled objectives proposed by theorists.

In conclusion, while the idea of population control through chemtrails taps into deeply rooted fears about governmental overreach and personal autonomy, it remains unsupported by empirical evidence. The persistence of this theory highlights the powerful influence of anecdotal evidence and the human tendency to seek explanations for complex social and health phenomena. As we continue to explore other alleged purposes of chemtrails, the importance of critical thinking and reliance on credible sources becomes increasingly evident. Understanding the motivations and psychological factors behind these beliefs is crucial for addressing and debunking the myths that sustain the chemtrails narrative.

Mind Control and Psychological Manipulation

Among the more alarming claims associated with the chemtrails conspiracy theory is the notion that these trails are part of a covert operation to exert mind control or psychological manipulation over the population. This theory suggests that chemtrails contain substances designed to influence human behavior, emotions, and cognitive functions, thus enabling governments or other powerful entities to maintain control over the populace.

Proponents of the mind control theory point to a variety of substances that they believe are being dispersed through chemtrails. These include psychotropic drugs, nano-particles, and neurotoxins, which are said to be capable of altering brain chemistry and affecting mental states. The alleged goal of this operation is to create a more compliant and manageable population, suppress dissent, and control societal behavior on a large scale.

The concept of using chemicals for mind control is not entirely new and has its roots in historical and contemporary examples of psychological manipulation. For instance, during the Cold War, both the United States and the Soviet Union conducted experiments on mind control and brainwashing techniques, often using drugs like LSD. These experiments, some of which were declassified in later years, provided a basis for the belief that governments have the capability and willingness to engage in psychological manipulation.

Modern-day proponents of the chemtrails mind control theory often cite these historical precedents, arguing that advancements in technology have made such operations more feasible and sophisticated. They claim that substances dispersed through chemtrails are specifically designed to target the central nervous system, causing changes in mood, cognitive function, and even physical health. These changes are said to manifest as increased anxiety, depression, lethargy, and a general sense of malaise, making the population more susceptible to control and less likely to resist authority.

However, these claims are met with significant skepticism from the scientific community. There is no credible evidence to support the idea that chemtrails contain mind-altering substances or that such an operation is technically feasible on the scale suggested by proponents. The atmospheric dispersion of chemicals would be highly unpredictable and inefficient, making it an unlikely method for targeted psychological manipulation.

Moreover, extensive air and water quality monitoring conducted by environmental agencies have found no traces of the substances alleged to be used for mind control. These findings align with the understanding that contrails are composed of water vapor and ice crystals, not exotic chemicals designed to influence human behavior.

The psychological appeal of the mind control theory can be partially explained by the human tendency to seek explanations for complex social and psychological phenomena. In times of uncertainty and anxiety, attributing personal and societal challenges to a covert, external force provides a sense of clarity and control. This explanation can be more comforting than confronting the multifaceted and often ambiguous nature of mental health issues and social dynamics.

In conclusion, while the idea of mind control through chemtrails is a captivating and alarming narrative, it remains unsupported by empirical evidence. The persistence of this theory highlights the powerful influence of historical precedents, anecdotal evidence, and the human need for simple explanations to complex problems. As we continue to explore other alleged purposes of chemtrails, the importance of critical thinking and reliance on credible sources becomes increasingly evident. Understanding the motivations and psychological factors behind these beliefs is crucial for addressing and debunking the myths that sustain the chemtrails narrative.

Environmental and Ecological Impact

Another angle from which chemtrails conspiracy theorists view these supposed operations is their impact on the environment and ecosystems. Proponents argue that the chemicals allegedly being dispersed from aircraft are designed to influence agricultural productivity, affect wildlife, and modify natural ecosystems. These claims add a layer of environmental activism to the theory, appealing to those concerned about the planet's health and sustainability.

One of the central claims is that chemtrails are being used to influence agricultural output. Proponents suggest that chemicals sprayed from planes are meant to enhance or inhibit crop growth, ostensibly to control food supply for geopolitical or economic purposes. Some allege that certain chemicals increase soil salinity, making it difficult for plants to absorb water and nutrients, thereby reducing agricultural yields. Others believe that these chemicals are used to engineer more resilient crop strains that can withstand extreme weather conditions, which may benefit agribusiness corporations at the expense of small farmers.

Wildlife is another area of concern for chemtrails theorists. They claim that the substances being released into the atmosphere are harmful to animals, particularly birds and bees. Declining bee populations, often referred to as colony collapse disorder, are cited as evidence of the detrimental effects of chemtrails. The theory posits that chemicals in the air disrupt the navigation systems of bees, leading to their inability to return to hives and ultimately causing the collapse of bee colonies. Similarly, bird deaths and unusual migratory patterns are attributed to chemical exposure from chemtrails, with proponents arguing that these chemicals interfere with birds' natural behaviors and health.

The broader ecological impact of chemtrails is also a significant part of the narrative. Proponents argue that the chemicals being sprayed affect entire ecosystems, leading to changes in soil composition, water quality, and plant health. They suggest that these changes can have cascading effects, altering the balance of local ecosystems and leading to the decline of certain species. For example, they claim that increased soil acidity caused by chemtrails can harm plants and aquatic life, disrupting food chains and ecosystems.

However, scientific evidence does not support these claims. Extensive environmental monitoring and research have found no unusual concentrations of the chemicals alleged to be used in chemtrails. Studies conducted by environmental agencies and independent researchers consistently show that the levels of heavy metals and other substances in soil, water, and air samples are within normal ranges and are not linked to aircraft emissions. Additionally, the observed declines in bee populations and other wildlife health issues can be attributed to well-documented factors such as pesticide use, habitat loss, and climate change.

The scientific community also points out the logistical and practical challenges of using chemtrails for environmental modification. Dispersion of chemicals from high altitudes would be highly unpredictable and inefficient, making it an unlikely method for targeted ecological engineering. Furthermore, the notion of a coordinated global program to manipulate ecosystems through chemtrails is implausible given the complexity and variability of natural environments.

In conclusion, while the environmental and ecological impact claims associated with chemtrails tap into genuine concerns about the planet's health, they remain unsupported by scientific evidence. The persistence of these theories highlights the intersection of environmental activism and conspiracy thinking, where legitimate concerns are co-opted into unfounded narratives. As we continue to explore other alleged purposes of chemtrails, the importance of critical thinking and reliance on credible sources becomes increasingly evident. Understanding the motivations and psychological factors behind these beliefs is crucial for addressing and debunking the myths that sustain the chemtrails narrative.

Other Alleged Purposes and Theories

Beyond the more commonly discussed claims of weather modification, population control, mind control, and ecological impact, the chemtrails conspiracy theory encompasses a variety of other alleged purposes that range from the plausible to the fantastical. These additional theories contribute to the rich tapestry of beliefs that sustain the overall narrative and illustrate the diverse ways in which people interpret and assign meaning to the phenomenon of persistent trails in the sky.

One of the less common but persistent theories is the idea that chemtrails are used for economic control and corporate interests. Proponents of this theory suggest that the chemicals allegedly being sprayed are intended to harm or alter the crops of small farmers, thereby driving them out of business and consolidating agricultural control in the hands of large agribusiness corporations. This theory posits that by manipulating environmental conditions, these corporations can create a dependency on genetically modified crops or proprietary seeds that are designed to withstand the chemical agents dispersed through chemtrails.

Another related theory involves the use of chemtrails for geopolitical strategy and warfare. Some believe that chemtrails are part of a covert program to induce droughts, floods, or other extreme weather events in specific regions, thereby destabilizing economies and weakening adversaries. This idea is rooted in the historical context of weather modification being considered as a tool of war, such as during the Vietnam War when the U.S. military used cloud seeding to extend the monsoon season. Although there is a basis in history for weather modification in warfare, there is no credible evidence to support the idea that chemtrails are currently being used for such purposes on a global scale.

A more fantastical theory suggests that chemtrails are involved in extraterrestrial communication or defense. Proponents of this idea believe that the substances being dispersed are meant to create atmospheric conditions conducive to communicating with or defending against alien life forms. This theory often intersects with broader narratives about government cover-ups of extraterrestrial encounters and advanced technologies. While intriguing, this claim lacks any empirical support and is generally considered to be on the fringes of the chemtrails narrative.

Misinformation and disinformation play a significant role in perpetuating these diverse theories. The rapid spread of information on the internet allows for the quick dissemination of unfounded claims and speculative ideas. Social media platforms, alternative media outlets, and personal blogs contribute to an echo chamber effect, where information that aligns with the chemtrails narrative is amplified and repeated, while contradictory evidence is ignored or dismissed. This creates a self-reinforcing cycle that sustains belief in the theory despite the lack of credible evidence.

Psychological and sociological factors also drive the belief in these varied purposes. Conspiracy theories often provide a sense of order and explanation in a world that can seem chaotic and unpredictable. They offer simple answers to complex problems and can be particularly appealing to individuals who feel disempowered or disconnected from mainstream institutions. The chemtrails theory, with its multiple alleged purposes, allows believers to find a narrative that resonates with their personal fears, suspicions, and worldviews.

In conclusion, the diversity of alleged purposes for chemtrails highlights the flexibility and adaptability of conspiracy theories. While some claims may seem more plausible than others, they all share a common foundation in mistrust of authority and a desire to find hidden meanings behind observable phenomena. As we continue to explore the world of chemtrails, understanding the psychological and sociological underpinnings of these beliefs is crucial for addressing and debunking the myths that sustain the narrative. Critical thinking and reliance on credible sources remain essential tools in disentangling fact from fiction in the complex landscape of modern conspiracy theories.

Chapter 5: Government and Military Involvement

Allegations of Government and Military Operations
The chemtrails conspiracy theory is deeply intertwined with allegations of secretive government and military operations. Proponents of the theory argue that chemtrails are part of covert projects run by various government agencies and branches of the military, designed to achieve a range of hidden objectives. These claims often point to shadowy, high-level collaborations involving national and international entities, suggesting a vast and coordinated effort to manipulate the environment and population.

Central to these allegations is the belief that the government has long been engaged in experimentation with atmospheric manipulation. Proponents argue that chemtrails are a continuation of historical weather modification programs and other classified projects. They claim that these operations are conducted with the knowledge and support of government agencies such as the Central Intelligence Agency (CIA), the Department of Defense (DoD), and even civilian organizations like the Environmental Protection Agency (EPA).

One of the key events often cited by theorists is the publication of the 1996 United States Air Force report titled "Weather as a Force

Multiplier: Owning the Weather in 2025." This document speculated on future technological advancements in weather modification and suggested potential military applications. Conspiracy theorists have seized upon this report as evidence that the military is actively pursuing such capabilities. Despite the speculative nature of the report and the absence of direct evidence linking it to chemtrail operations, it has become a cornerstone of the chemtrails narrative.

Theorists also point to specific documents and government programs as proof of chemtrail activities. Declassified projects like Project Popeye, a weather modification program conducted during the Vietnam War, are frequently mentioned. Project Popeye involved cloud seeding to extend the monsoon season and disrupt enemy supply lines. Although it is a well-documented historical example, theorists extrapolate from it to argue that similar, more advanced operations are currently underway, involving the dispersion of chemicals from commercial and military aircraft.

Additionally, proponents often reference patents related to atmospheric and weather modification technologies. These patents, filed by various inventors and companies, describe methods for cloud seeding, solar radiation management, and other geoengineering techniques. While the existence of these patents demonstrates interest in atmospheric manipulation, they do not provide concrete evidence of active chemtrail operations. Nonetheless, they are cited by theorists as part of a broader pattern of governmental and corporate interest in controlling the environment.

The scale of the alleged operations is another critical aspect of these claims. Theorists argue that the sheer number of aircraft involved, coupled with the need for secrecy and coordination, implies a high level of governmental and military involvement. They suggest that the operations are conducted globally, with commercial airlines and military planes being outfitted with specialized equipment to disperse chemicals. This notion of a vast, coordinated effort aligns with broader conspiracy theories that posit the existence of powerful, clandestine groups manipulating global events.

Despite the elaborate nature of these claims, they face significant scrutiny from the scientific community and government officials. Skeptics argue that the logistical challenges and lack of credible evidence make the theory highly implausible. The scientific explanations for contrail formation and persistence are well established, and extensive air quality testing has not detected the presence of unusual or harmful substances attributed to chemtrails.

In conclusion, while the allegations of government and military involvement in chemtrail operations are a central component of the conspiracy theory, they remain unsupported by concrete evidence. These claims reflect broader themes of mistrust and suspicion towards authority, tapping into historical examples and speculative documents to construct a compelling, albeit unfounded, narrative. As we continue to explore the chemtrails theory, it is essential to critically evaluate the sources of information and the motivations behind these allegations.

Declassified Documents and Their Interpretations

The world of conspiracy theories is often built upon the reinterpretation of declassified documents, and the chemtrails narrative is no exception. Proponents of the chemtrails theory frequently cite a range of declassified government documents as evidence supporting their claims. These documents, often related to atmospheric and weather modification programs, are reinterpreted through a conspiratorial lens to fit the narrative that chemtrails are part of a covert operation.

One of the most frequently referenced documents in the chemtrails community is the 1996 United States Air Force report titled "Weather as a Force Multiplier: Owning the Weather in 2025." This report, which was purely speculative and intended to explore potential future technological advancements, discussed various methods for weather modification. It suggested that by 2025, it might be possible to use weather modification as a military tool to gain strategic advantages. Although the report was theoretical, conspiracy theorists have interpreted it as proof that the military is already engaging in large-scale weather modification through chemtrails.

In addition to the USAF report, conspiracy theorists often point to declassified documents from historical projects such as Project Popeye. During the Vietnam War, the U.S. military conducted this cloud seeding operation with the aim of extending the monsoon season to disrupt enemy supply lines. The project involved dispersing silver iodide into the atmosphere to induce rainfall. While Project Popeye is well-documented and widely acknowledged, proponents of the chemtrails theory extrapolate from this historical example to argue that similar, more advanced operations are currently underway, involving the dispersion of chemicals from commercial and military aircraft.

Another set of documents often cited by chemtrails theorists includes patents related to weather modification and geoengineering technologies. These patents describe various methods for cloud seeding, solar radiation management, and other atmospheric interventions. While the existence of these patents demonstrates ongoing interest and research in atmospheric modification, they are frequently misinterpreted or taken out of context to support the idea that chemtrails are a current and active operation.

The reinterpreted documents extend beyond just weather modification to include various government research projects on atmospheric testing and environmental monitoring. For example, documents related to the High Frequency Active Auroral Research Program (HAARP), a research program jointly funded by the U.S.

Air Force, Navy, and the Defense Advanced Research Projects Agency (DARPA), are often cited. HAARP was designed to study the ionosphere and develop enhancement technologies for radio communications and surveillance. Despite its legitimate scientific goals, HAARP has been the subject of numerous conspiracy theories, including those related to chemtrails, with proponents claiming it is used for mind control or weather manipulation.

Despite the frequent citation of these documents, the scientific community and government officials have consistently debunked the interpretations put forth by chemtrails proponents. Declassified documents and patents related to atmospheric science are often publicly available and thoroughly analyzed. Experts point out that these documents typically describe theoretical research, historical projects, or speculative technologies rather than active, covert operations. Furthermore, the logistics and feasibility of conducting global chemical spraying operations on the scale suggested by chemtrails theorists are highly implausible.

In summary, the reinterpretation of declassified documents plays a significant role in sustaining the chemtrails conspiracy theory. Proponents selectively highlight and misinterpret these documents to construct a narrative that aligns with their beliefs. However, a closer examination of the actual content and context of these documents reveals that they do not support the existence of a covert chemtrails operation. As we continue to explore the chemtrails theory, it is crucial to critically evaluate the sources of information and understand the motivations behind these reinterpretations.

Public Statements and Denials from Officials

From the outset, public statements and official denials have played a significant role in the discourse surrounding the chemtrails conspiracy theory. Government agencies, military officials, and scientists have repeatedly addressed the claims, seeking to debunk the theory and reassure the public that there is no covert operation involving chemical spraying from aircraft. Despite these efforts, the mistrust and skepticism from conspiracy theorists persist, fueling ongoing debate and speculation.

One of the most notable instances of official denial came from the United States Air Force (USAF). In response to increasing public inquiries and concerns, the USAF published a fact sheet in 2000 titled "Chemtrails—Fact or Fiction?" This document aimed to clarify the nature of contrails and address the misconceptions perpetuated by the chemtrails theory. The fact sheet provided a detailed explanation of contrail formation, emphasizing that these trails are composed of water vapor and ice crystals resulting from aircraft engine exhaust. The USAF categorically denied any involvement in chemical spraying operations, stating that there was no evidence to support the existence of chemtrails.

Similarly, NASA and the Environmental Protection Agency (EPA) have issued statements debunking the chemtrails theory. NASA has consistently explained that contrails are a natural byproduct of high-altitude flight, formed when hot, humid exhaust gases from jet engines mix with the cold air of the upper atmosphere. The EPA has conducted numerous studies on air quality and found no evidence of unusual or harmful substances associated with contrails. These agencies have sought to educate the public through informational campaigns, scientific reports, and collaborations with other scientific organizations.

Despite these efforts, public reception of official statements has been mixed. For many conspiracy theorists, the denials and explanations offered by government agencies are seen as part of the cover-up. The mistrust of authority and the belief in hidden agendas lead some individuals to interpret these statements as attempts to conceal the truth. This dynamic creates a feedback loop where official denials reinforce the perception of a conspiracy rather than dispelling it.

Case studies of official responses provide insight into the complexities of addressing conspiracy theories. For example, in 2016, the United Nations hosted a high-level meeting on climate change, where the issue of geoengineering was discussed. Following the meeting, several conspiracy theorists claimed that chemtrails were part of the agenda, despite no evidence to support such claims. Official statements from the UN clarified that the meeting focused on legitimate scientific research into climate change mitigation, but these denials did little to quell the speculations among believers.

Another example is the response from the German government. In 2013, a member of the German Bundestag raised the issue of chemtrails in parliament, prompting an official investigation. The government concluded that there was no evidence to support the existence of chemtrails and that contrails were a well-understood phenomenon. Despite this clear and public denial, conspiracy theorists in Germany continued to promote the chemtrails narrative, interpreting the government's response as insufficient or misleading.

The persistent mistrust in official explanations highlights the challenges faced by authorities in combating conspiracy theories. Efforts to debunk such theories through scientific evidence and transparent communication are often met with skepticism and resistance. This dynamic underscores the importance of building public trust and fostering critical thinking skills to effectively address misinformation.

In conclusion, public statements and denials from officials have been a key aspect of the ongoing discourse around chemtrails. While these efforts aim to clarify misconceptions and provide accurate information, they often face significant obstacles in overcoming deep-seated mistrust and skepticism. Understanding this interplay between official narratives and conspiracy beliefs is crucial for developing strategies to address and mitigate the impact of misinformation.

Whistleblowers and Insider Testimonies

The chemtrails conspiracy theory owes much of its allure and persistence to the accounts of self-proclaimed whistleblowers and insiders who claim to have firsthand knowledge of secretive government and military operations. These individuals often present dramatic testimonies that purport to expose clandestine activities involving the dispersion of chemicals from aircraft. While their stories are compelling, they are frequently lacking in verifiable evidence and are often met with skepticism from the scientific community and government authorities.

One of the most notable figures in this realm is Clifford Carnicom, a former government employee who has devoted considerable effort to researching and promoting the chemtrails theory. Carnicom's technical background lends his claims an air of credibility, and his detailed reports and presentations have made him a prominent voice within the chemtrails community. He alleges that the government is conducting large-scale geoengineering projects involving the release of harmful chemicals into the atmosphere, though his findings have not been corroborated by independent sources.

Another significant whistleblower is Kristen Meghan, a former industrial hygienist for the U.S. Air Force. Meghan has publicly claimed that she uncovered evidence of large quantities of chemicals being procured and used in operations related to chemtrails during her tenure with the Air Force. Her articulate presentations and professional background have made her a persuasive spokesperson for the theory, though her allegations have not been substantiated by independent investigations.

Dr. Bill Deagle, a former medical doctor, is another key figure who has gained attention for his dramatic claims. Deagle asserts that he has insider knowledge of various government programs, including those related to weather modification and chemical spraying. His presentations are known for their detailed and often sensational accounts, which have captivated audiences at conferences and in online videos. However, Deagle's lack of verifiable credentials and the extreme nature of some of his claims have led many to question his credibility.

These whistleblowers and insiders often present their testimonies through alternative media outlets, online videos, and conspiracy theory conferences. Their accounts typically share several common elements: descriptions of secretive programs, personal risk in coming forward, and anecdotal evidence rather than concrete proof. Despite the lack of verifiable evidence, these stories resonate with believers because they align with preexisting suspicions about government secrecy and malfeasance. The human element in these narratives makes the theory more relatable and emotionally impactful for those who are already inclined to believe.

The influence of these whistleblowers extends beyond individual testimonies. They often serve as catalysts for further speculation and investigation within the chemtrails community. Their accounts are frequently cited as "evidence" in online discussions and documentaries, reinforcing the belief that there is a hidden truth being concealed by official narratives. This dynamic creates a self-reinforcing cycle, where the testimonies of whistleblowers are used to validate the broader conspiracy theory, which in turn attracts more individuals willing to come forward with similar claims.

However, the scientific community and government officials continue to challenge these testimonies. They point out that the logistical challenges and lack of concrete evidence make the theory highly implausible. Extensive air quality testing and environmental monitoring have not detected the presence of unusual or harmful substances attributed to chemtrails. The explanations provided by whistleblowers often fail to account for the established scientific understanding of contrail formation and atmospheric science.

In conclusion, while the testimonies of whistleblowers and insiders are a central component of the chemtrails conspiracy theory, they remain unsupported by verifiable evidence. These accounts reflect broader themes of mistrust and suspicion towards authority, tapping into historical examples and speculative documents to construct a compelling, albeit unfounded, narrative. As we continue to explore the chemtrails theory, it is essential to critically evaluate the sources of information and the motivations behind these allegations. Understanding the psychological and sociological factors that drive belief in such testimonies is crucial for addressing and debunking the myths that sustain the chemtrails narrative.

International Perspectives and Collaborations

While much of the discourse around chemtrails centers on alleged operations within the United States, the theory extends well beyond American borders, encompassing international perspectives and supposed collaborations between governments and military alliances. This global dimension adds a layer of complexity to the narrative, suggesting a coordinated effort among various nations to carry out atmospheric and environmental manipulation.

Proponents of the chemtrails theory argue that the phenomenon is not confined to any one country but is part of a worldwide agenda involving multiple governments and international organizations. They posit that chemtrails are being used to address global issues such as climate change, food security, and population control. According to these claims, international bodies such as the United Nations, NATO, and other transnational organizations are complicit in these operations, coordinating efforts across borders.

One of the key pieces of "evidence" cited by theorists is the existence of international agreements and treaties related to environmental and atmospheric activities. For instance, the 1978 Environmental Modification Convention (ENMOD), which prohibits the military or other hostile use of environmental modification techniques, is often mentioned. Theorists argue that the very existence of such a treaty implies that countries have developed and possibly deployed environmental modification technologies. However, the ENMOD treaty's primary purpose is to prevent environmental warfare, and there is no credible evidence to suggest that it supports the existence of chemtrails.

Additionally, documents and statements from international organizations are frequently reinterpreted to fit the chemtrails narrative. For example, discussions about geoengineering and climate intervention by reputable scientific bodies are often taken out of context. Proponents claim that these discussions are thinly veiled admissions of ongoing chemtrail activities, despite the fact that geoengineering research is typically focused on potential future solutions to mitigate climate change and is conducted transparently with public oversight.

The global spread of the chemtrails theory is facilitated by the internet and social media, which allow information and misinformation to cross borders effortlessly. Websites, forums, and social media groups dedicated to chemtrails have international memberships, with users from various countries sharing their observations and interpretations. This digital connectivity has enabled the theory to gain traction in regions far removed from its origins in the United States.

In countries like Canada, the United Kingdom, Germany, and Australia, local variations of the chemtrails theory have emerged, tailored to specific cultural and political contexts. For instance, in Germany, some proponents link chemtrails to broader concerns about environmental pollution and government transparency. In Australia, the theory intersects with anxieties about droughts and water scarcity. These localized versions of the theory often incorporate unique elements while maintaining the core belief in a global chemtrails agenda.

Despite the international reach of the chemtrails theory, it remains unsupported by credible scientific evidence. Atmospheric scientists and environmental researchers around the world have consistently debunked the claims, emphasizing that contrails are a well-understood phenomenon resulting from aircraft engine exhaust in cold, humid conditions. Air quality and environmental monitoring in various countries have found no unusual or harmful substances that would indicate chemical spraying operations.

In conclusion, the international perspectives and collaborations alleged by chemtrails proponents add a global dimension to the theory, suggesting a coordinated effort among multiple governments and organizations. While the existence of international treaties and discussions about geoengineering are reinterpreted to fit the narrative, there is no credible evidence to support these claims. The persistence of the chemtrails theory highlights the challenges of combating misinformation on a global scale and underscores the importance of critical thinking and reliance on scientific expertise. As we continue to explore the world of chemtrails, understanding the broader context and motivations behind these beliefs is crucial for addressing and debunking the myths that sustain the narrative.

Chapter 6: The Role of Whistleblowers and Insiders

Accounts from Alleged Whistleblowers

The chemtrails conspiracy theory owes much of its traction to the accounts of self-proclaimed whistleblowers who claim insider knowledge about these alleged operations. These individuals, often presenting themselves as former government or military personnel, scientists, or aviation experts, assert that they have firsthand experience with secretive programs involving the dispersion of chemicals from aircraft. Their testimonies, filled with dramatic details and purported evidence, have significantly influenced public perception and fueled the persistence of the chemtrails narrative.

One of the earliest and most influential figures in this context is Clifford Carnicom. A former government employee, Carnicom became a prominent voice in the chemtrails community by dedicating his efforts to researching and promoting the theory. He produced a series of reports and presentations that detailed his findings, which he claimed were based on scientific analysis and insider information. Carnicom's technical background lent his claims an air of credibility, and his work has been widely cited by proponents of the chemtrails theory.

Another notable whistleblower is Kristen Meghan, a former industrial hygienist and environmental specialist for the U.S. Air Force. Meghan has publicly claimed that during her tenure with the Air Force, she uncovered evidence of large quantities of chemicals being procured and used in operations related to chemtrails. Her articulate presentations and professional background have made her a compelling spokesperson for the theory. Meghan's claims have been featured in numerous documentaries, interviews, and alternative media platforms, further amplifying her message.

Dr. Bill Deagle, a former medical doctor, has also gained attention for his dramatic and detailed allegations. Deagle asserts that he has insider knowledge of various government programs, including those related to weather modification and chemical spraying. His presentations, often delivered with a sense of urgency and authority, have captivated audiences at conferences and in online videos. However, his lack of verifiable credentials and the extreme nature of some of his claims have led many to question his credibility.

These whistleblowers and insiders often describe secretive programs involving the dispersion of chemicals from both military and commercial aircraft. They claim that these operations are conducted with the knowledge and support of high-level government and military officials, and that the true purpose of chemtrails is being concealed from the public. The substances allegedly dispersed, according to these accounts, include heavy metals, biological agents, and other toxic chemicals designed to achieve various covert objectives.

The impact of these testimonies on public perception cannot be overstated. For many believers, the accounts of whistleblowers serve as compelling evidence that corroborates their suspicions about chemtrails. The detailed and often emotional nature of these stories makes them particularly persuasive, creating a sense of urgency and legitimacy around the chemtrails narrative. These testimonies are widely shared on social media and alternative media platforms, where they reach a large audience and further entrench the belief in chemtrails.

However, the scientific community and government officials have consistently challenged these accounts. They point out that the logistical challenges and lack of concrete evidence make the theory highly implausible. Extensive air quality testing and environmental monitoring have not detected the presence of unusual or harmful substances attributed to chemtrails. The explanations provided by whistleblowers often fail to account for the established scientific understanding of contrail formation and atmospheric science.

In conclusion, the accounts of alleged whistleblowers play a central role in sustaining the chemtrails conspiracy theory. These testimonies, filled with dramatic details and purported insider knowledge, have significantly influenced public perception and belief in chemtrails. As we continue to explore the chemtrails theory, it is essential to critically evaluate the sources of information and the motivations behind these accounts. Understanding the psychological and sociological factors that drive belief in such testimonies is crucial for addressing and debunking the myths that sustain the narrative.

Analysis of Key Testimonies

The chemtrails conspiracy theory is bolstered by the testimonies of various self-proclaimed whistleblowers who claim to have insider knowledge about covert operations. These individuals present their stories with compelling details and assert that they possess direct evidence of chemical spraying activities. A critical analysis of these key testimonies is essential to understand their influence and assess their credibility.

One of the most prominent testimonies comes from Clifford Carnicom. Carnicom, who presents himself as a former government scientist, has produced numerous reports and videos that claim to expose the truth about chemtrails. He asserts that the trails left by aircraft are laden with harmful substances designed for geoengineering and population control. Carnicom's work includes detailed analyses of atmospheric samples, which he claims contain high levels of aluminum, barium, and other metals. However, independent scientific reviews of his methods and findings have raised significant doubts about their validity. Experts point out that Carnicom's sampling techniques lack rigorous scientific standards, and his conclusions are not supported by peer-reviewed research.

Kristen Meghan's testimony is another cornerstone of the chemtrails narrative. As a former industrial hygienist for the U.S. Air Force, Meghan claims that she discovered evidence of large-scale chemical procurement and deployment during her service. She alleges that she witnessed the purchase of chemicals in quantities that far exceeded typical requirements, suggesting their use in undisclosed projects. Meghan's assertions have been featured in multiple documentaries and interviews, lending her claims a degree of visibility and influence. However, her evidence primarily consists of anecdotal accounts and unverified documentation, which have not been substantiated by independent investigations. Critics argue that her interpretations of procurement records are speculative and do not conclusively link the chemicals to chemtrail activities.

Dr. Bill Deagle's testimony adds a dramatic element to the chemtrails theory. Deagle, who identifies himself as a former medical doctor with insider knowledge, claims that he has been privy to secret government programs involving chemical and biological agents. His presentations often include detailed descriptions of supposed chemical compositions and their intended effects on human populations. Deagle's claims are characterized by their sensational nature, which appeals to a certain segment of the conspiracy community. However, his lack of verifiable credentials and the absence of corroborating evidence undermine the credibility of his testimony. Independent researchers and skeptics have consistently challenged the plausibility of his assertions.

These testimonies, while varied in detail and delivery, share common themes: they describe secretive operations, emphasize the personal risk taken by the whistleblowers, and present anecdotal evidence rather than concrete proof. The impact of these testimonies is amplified by their circulation on social media and alternative media platforms, where they are often taken at face value and shared widely among believers.

Comparing the statements of these whistleblowers with available scientific and governmental data reveals significant discrepancies. For example, extensive air quality testing conducted by environmental agencies has found no evidence of unusual or harmful substances in areas where chemtrails are alleged to occur. The established scientific understanding of contrail formation—resulting from the condensation of water vapor in aircraft exhaust—provides a clear, non-conspiratorial explanation for the observed phenomena.

In conclusion, while the testimonies of key whistleblowers play a central role in sustaining the chemtrails conspiracy theory, they are fraught with inconsistencies and lack credible evidence. A critical examination of their claims reveals significant gaps and methodological flaws that undermine their validity. As we continue to explore the chemtrails narrative, it is essential to rely on rigorous scientific analysis and credible sources to separate fact from fiction. Understanding the influence of these testimonies and their appeal to the conspiracy-minded is crucial for addressing and debunking the myths that perpetuate the chemtrails theory.

Motivations and Credibility of Whistleblowers

Understanding the motivations and credibility of self-proclaimed whistleblowers is crucial in assessing the validity of their claims about chemtrails. These individuals often present themselves as brave truth-tellers exposing hidden dangers, but their backgrounds, motivations, and the evidence they provide must be critically examined to determine their reliability.

Firstly, the motivations behind whistleblowers coming forward with claims about chemtrails can vary widely. Some may be driven by genuine belief in their cause and a desire to protect public health and the environment. For instance, Clifford Carnicom, a former government employee, has dedicated significant time and effort to researching and promoting the chemtrails theory. His motivations appear to stem from a conviction that he is uncovering important truths and safeguarding humanity. Similarly, Kristen Meghan, a former industrial hygienist for the U.S. Air Force, often frames her testimony as an act of public service, aiming to inform and protect the public from perceived harm.

However, motivations can also include personal gain, such as financial profit, fame, or a sense of importance. Some whistleblowers have monetized their claims through speaking engagements, books, and documentaries. Dr. Bill Deagle, for example, has become a notable figure within the chemtrails community, attracting attention and gaining a platform to promote his ideas. The potential for personal gain can sometimes cloud the objectivity of their testimonies, making it important to scrutinize the evidence they provide critically.

The personal and professional backgrounds of these individuals also play a significant role in evaluating their credibility. For example, while Clifford Carnicom presents himself as a former government scientist, independent investigations into his qualifications and previous work experience are necessary to verify his claims. Similarly, Kristen Meghan's background as an industrial hygienist lends some credibility to her assertions, but her lack of independent corroboration weakens the overall reliability of her testimony.

Another aspect to consider is the consistency of their statements. Credible whistleblowers typically provide consistent, detailed accounts that can be corroborated by independent evidence. In contrast, some of the claims made by chemtrails whistleblowers are often characterized by sensationalism and lack of verifiable data. Dr. Bill Deagle's presentations, for instance, are filled with dramatic allegations but lack the concrete evidence needed to support such claims. The extreme nature of some of his statements also raises questions about their plausibility.

Public and media response to whistleblower testimonies can significantly impact their perceived credibility. Whistleblowers like Kristen Meghan have gained a following through appearances in documentaries and interviews, which amplifies their message and lends a veneer of legitimacy. However, the absence of rigorous journalistic standards in some of these media outlets means that their claims are often presented without adequate scrutiny. This uncritical amplification can contribute to the persistence of the chemtrails narrative despite the lack of credible evidence.

Moreover, the influence of confirmation bias cannot be underestimated. Believers in the chemtrails theory are more likely to accept and amplify the testimonies of whistleblowers that align with their preexisting beliefs. This psychological phenomenon reinforces the belief in chemtrails and makes it challenging to introduce counterarguments or debunking evidence.

In conclusion, evaluating the motivations and credibility of chemtrails whistleblowers requires a thorough examination of their backgrounds, the consistency and plausibility of their claims, and the potential for personal gain. While some individuals may be genuinely convinced of their assertions, the lack of independent corroboration and the sensational nature of many testimonies undermine their reliability. Understanding these factors is crucial for critically assessing the chemtrails narrative and addressing the myths that sustain it. As we continue to explore the role of whistleblowers in the chemtrails theory, it is essential to rely on credible sources and rigorous analysis to separate fact from fiction.

Influence on Public Perception and Belief

The testimonies of self-proclaimed whistleblowers and insiders have profoundly shaped public perception and belief in the chemtrails conspiracy theory. These accounts, often detailed and dramatic, provide a seemingly credible foundation for the theory and have significantly influenced how it is received by the general public and conspiracy theory enthusiasts alike.

Whistleblower testimonies are frequently disseminated through social media platforms, alternative media outlets, and online forums dedicated to conspiracy theories. These platforms serve as echo chambers where like-minded individuals share and reinforce their beliefs. Videos, interviews, and written accounts from whistleblowers are shared widely, often accompanied by emotional and urgent appeals to the public to "wake up" and recognize the alleged truth about chemtrails. This widespread dissemination ensures that the narrative reaches a large audience, many of whom may already be predisposed to distrust governmental and scientific authority.

Social media plays a crucial role in amplifying the reach and impact of whistleblower testimonies. Platforms like Facebook, YouTube, and Twitter allow users to share content with a global audience instantaneously. Videos featuring whistleblower accounts often go viral, attracting millions of views and generating extensive discussion. The visual and emotional appeal of these videos makes them particularly effective at capturing viewers' attention and persuading them of the validity of the claims being made. Comment sections and discussion threads provide a space for believers to exchange experiences and interpretations, further reinforcing their shared beliefs.

Alternative media outlets, such as conspiracy theory websites and independent news platforms, also play a significant role in shaping public perception. These outlets often present whistleblower testimonies as breaking news or exclusive revelations, lending an air of legitimacy to the claims. The lack of editorial oversight and fact-checking in many of these platforms means that the testimonies are often presented without critical examination, allowing misinformation to spread unchecked. Articles and videos on these sites frequently frame whistleblower accounts within a broader context of governmental malfeasance and secrecy, resonating with readers who are already skeptical of official narratives.

Case studies of specific whistleblower testimonies illustrate their profound impact on public belief in chemtrails. For example, Kristen Meghan's claims about chemical procurement and deployment during her time with the U.S. Air Force have been widely circulated and cited as credible evidence by chemtrails proponents. Her professional background and articulate presentations have made her a persuasive spokesperson for the theory, and her interviews and videos have garnered significant attention. Similarly, Clifford Carnicom's detailed reports and scientific-sounding analyses have provided a veneer of legitimacy to the chemtrails narrative, further entrenching it in the public consciousness.

The broader impact of whistleblower accounts on the persistence of the chemtrails conspiracy theory is substantial. These testimonies create a sense of urgency and legitimacy around the theory, making it more difficult for debunking efforts to gain traction. The personal risk and sacrifice often emphasized by whistleblowers in their stories lend a sense of heroism and authenticity to their claims, making them more relatable and compelling to the public. As a result, the chemtrails narrative continues to thrive, fueled by the persuasive power of these insider accounts.

In conclusion, whistleblower testimonies have a profound influence on public perception and belief in the chemtrails conspiracy theory. Their accounts, widely disseminated through social media and alternative media outlets, provide a seemingly credible foundation for the theory and significantly shape how it is received by the general public. The emotional and detailed nature of these testimonies makes them particularly persuasive, reinforcing the belief in chemtrails and challenging efforts to debunk the theory. Understanding the impact of these testimonies is crucial for addressing and countering the myths that sustain the chemtrails narrative.

Skepticism and Counterarguments

The testimonies of self-proclaimed whistleblowers and insiders have undoubtedly fueled the chemtrails conspiracy theory, but they have also attracted significant skepticism and counterarguments from the scientific community, government agencies, and critical thinkers. Understanding these counterarguments is crucial for evaluating the validity of whistleblower claims and addressing the persistent myths surrounding chemtrails.

Skeptics often point to the lack of concrete evidence and the methodological flaws in whistleblower testimonies. For example, Clifford Carnicom's atmospheric samples and analyses, frequently cited by chemtrails proponents, have been criticized for their lack of scientific rigor. Experts argue that Carnicom's sampling methods do not adhere to established scientific standards, and his conclusions are not supported by peer-reviewed research. Independent studies have consistently found that contrails are composed of water vapor and ice crystals, not harmful chemicals.

Similarly, Kristen Meghan's claims about chemical procurement and deployment have faced scrutiny. Critics argue that her interpretations of procurement records are speculative and do not provide definitive proof of chemtrail activities. Environmental agencies, such as the Environmental Protection Agency (EPA) and the Federal Aviation Administration (FAA), have conducted extensive air quality testing and found no evidence of unusual or harmful substances associated with contrails. These findings challenge Meghan's assertions and underscore the importance of relying on verifiable data.

Dr. Bill Deagle's dramatic allegations have also been met with skepticism. His lack of verifiable credentials and the sensational nature of his claims undermine their credibility. Skeptics point out that Deagle's presentations often lack specific, corroborated evidence and instead rely on fear-mongering and anecdotal accounts. The scientific community emphasizes the need for concrete, reproducible evidence to support such extraordinary claims.

One of the primary counterarguments to the chemtrails theory is the logistical and practical challenges of conducting a global chemical spraying operation. The theory posits that thousands of aircraft, both commercial and military, are involved in dispersing chemicals across the globe. However, skeptics argue that coordinating such an extensive operation would require immense resources, secrecy, and cooperation among various governments and agencies—a scenario that is highly implausible given the complexity and transparency of modern aviation.

Furthermore, skeptics highlight the well-documented science behind contrail formation. Contrails are formed when hot, humid exhaust gases from aircraft engines mix with the cold, low-pressure air of the upper atmosphere, causing the water vapor to condense and freeze into ice crystals. The variability in contrail appearance and persistence is explained by atmospheric conditions such as temperature, humidity, and wind patterns. This scientific understanding is supported by decades of research and observation, providing a clear, non-conspiratorial explanation for the observed phenomena.

The role of independent fact-checking organizations and science communicators is also crucial in countering the chemtrails narrative. These organizations work to debunk misinformation and provide accurate information to the public. For example, websites like Snopes and FactCheck.org have published detailed articles addressing the chemtrails theory, explaining the science behind contrails and debunking common myths. Science communicators, such as meteorologists and aviation experts, often engage with the public through social media, blogs, and public lectures to clarify misconceptions and promote scientific literacy.

In conclusion, while whistleblower testimonies have played a significant role in sustaining the chemtrails conspiracy theory, they face substantial skepticism and counterarguments from the scientific community and critical thinkers. The lack of concrete evidence, methodological flaws, and logistical challenges undermine the credibility of these claims. By understanding and addressing these counterarguments, we can better evaluate the validity of the chemtrails narrative and promote a more informed and scientifically literate public. As we continue to explore the chemtrails theory, it is essential to rely on credible sources and rigorous analysis to separate fact from fiction.

Chapter 7: Scientific Rebuttals and Debunking

Scientific Studies and Expert Opinions

The chemtrails conspiracy theory has been met with significant scrutiny from the scientific community, which has undertaken numerous studies to investigate and debunk the claims. These studies, conducted by atmospheric scientists, meteorologists, and environmental researchers, have consistently found no evidence to support the notion that aircraft contrails are anything other than condensed water vapor and ice crystals. The body of scientific literature addressing this topic is extensive, providing robust explanations for the phenomena often misinterpreted as chemtrails.

One of the foundational studies in this area was conducted by the Environmental Protection Agency (EPA) in response to growing public concern about chemtrails. The EPA's comprehensive air quality testing involved collecting and analyzing samples from areas where chemtrails were allegedly prevalent. The results consistently showed that the levels of chemicals like aluminum and barium were within normal environmental ranges and not indicative of any unusual atmospheric spraying. These findings aligned with the scientific understanding that contrails are composed of water vapor and do not contain harmful substances.

Similarly, NASA has played a crucial role in debunking the chemtrails theory through its research on atmospheric conditions and contrail formation. NASA scientists have used satellite data and high-altitude aircraft to study the formation and behavior of contrails. Their research has demonstrated that contrails form when hot, humid exhaust from jet engines meets the cold, low-pressure air of the upper atmosphere. This process causes the water vapor to condense and freeze, creating the visible trails. NASA's findings have been published in peer-reviewed journals, providing a transparent and scientifically rigorous explanation for contrail phenomena.

Meteorologists have also contributed significantly to the debunking efforts. Professionals in this field have conducted observational studies and computer simulations to understand the conditions under which contrails form and persist. Their work has shown that the appearance and duration of contrails are influenced by factors such as temperature, humidity, and wind patterns at high altitudes. Persistent contrails, which are often cited by chemtrail proponents as evidence of chemical spraying, can occur naturally when the atmospheric conditions are right. These scientific explanations help to clarify why some contrails linger and spread, creating cloud-like formations.

Expert opinions from atmospheric scientists and environmental researchers further bolster the scientific consensus against the chemtrails theory. Dr. David Travis, a professor of geography and an expert in atmospheric science, has extensively studied the impact of contrails on climate and weather. His research indicates that contrails can affect local climate by reflecting sunlight and trapping heat, but there is no evidence to suggest that they contain harmful chemicals. Dr. Travis and other experts emphasize the importance of relying on peer-reviewed research and credible scientific sources when evaluating claims about chemtrails.

In summary, the scientific studies and expert opinions on contrails provide a clear and consistent rebuttal to the chemtrails conspiracy theory. Through rigorous testing, observational research, and peer-reviewed publications, scientists have demonstrated that contrails are a natural byproduct of high-altitude flight and do not pose a threat to public health or the environment. These findings underscore the importance of scientific literacy and critical thinking in dispelling misinformation and understanding the natural phenomena that shape our world. As we continue to explore the chemtrails narrative, it is essential to rely on credible scientific evidence to separate fact from fiction and promote a more informed public discourse.

Common Misconceptions and How They Are Addressed

The chemtrails conspiracy theory is rife with misconceptions that have fueled its persistence and spread. To effectively debunk these claims, it is essential to address and correct the common misunderstandings that underlie them. By providing clear scientific explanations for the phenomena often misinterpreted as chemtrails, we can demystify these misconceptions and promote a more accurate understanding of atmospheric science.

One of the most pervasive misconceptions is the belief that the persistence and spread of contrails are indicative of chemical spraying. Proponents of the chemtrails theory argue that normal contrails should dissipate quickly, and that trails lingering for extended periods must be the result of chemical additives. However, scientific research has shown that the duration and appearance of contrails depend on atmospheric conditions such as temperature, humidity, and wind patterns. In high humidity, ice crystals in contrails can persist and spread, forming cirrus clouds that can last for hours. This natural variability explains why some contrails appear to linger longer than others, without invoking the need for chemical explanations.

Another common misconception is the interpretation of unusual cloud formations as evidence of chemtrails. Chemtrails theorists often point to cloud patterns such as crisscrossing lines, grid-like formations, and halos around the sun as signs of deliberate chemical spraying. These patterns, however, can be explained by well-understood atmospheric processes. For instance, wind shear at high altitudes can cause contrails to spread and intersect, creating the appearance of a grid. Halos around the sun are a natural optical phenomenon caused by the refraction of light through ice crystals in the atmosphere. These scientific explanations provide a clear, non-conspiratorial understanding of the observed phenomena.

Health concerns are another area rife with misconceptions. Chemtrails proponents claim that the substances allegedly dispersed from aircraft are causing a range of health issues, from respiratory problems to neurological disorders. They often cite anecdotal reports and spikes in illness as evidence of chemical exposure. However, extensive air quality monitoring and epidemiological studies have found no link between contrails and adverse health effects. The substances typically mentioned, such as aluminum and barium, are present in the environment at levels that are not harmful and are not associated with aircraft emissions. Public health agencies and environmental researchers emphasize that the observed health issues can be attributed to other factors, such as pollution, allergens, and lifestyle choices.

Another frequent claim is that government patents on weather modification and geoengineering technologies are proof of chemtrails operations. Chemtrails theorists often reference patents related to cloud seeding, solar radiation management, and other atmospheric interventions as evidence that these technologies are being actively used. While these patents exist and demonstrate interest in atmospheric modification, they do not provide proof of ongoing chemtrails activities. Patents represent potential technological developments and research interests, not necessarily active programs. Moreover, legitimate scientific research on geoengineering is conducted transparently and with public oversight, unlike the secretive operations alleged by chemtrails theorists.

Education and public awareness play crucial roles in addressing these misconceptions. By promoting scientific literacy and critical thinking, we can help the public understand the natural explanations for contrail formation and atmospheric phenomena. Outreach efforts by scientists, educators, and science communicators are essential for dispelling myths and providing accurate information. Engaging with the public through lectures, social media, and educational programs can counteract the spread of misinformation and build trust in credible sources.

In conclusion, addressing common misconceptions about contrails and chemtrails is vital for debunking the conspiracy theory. By providing clear, scientifically grounded explanations for the observed phenomena, we can demystify these claims and promote a more accurate understanding of atmospheric science. Education and public awareness efforts are essential for dispelling myths, promoting critical thinking, and fostering a scientifically literate society. As we continue to explore the chemtrails narrative, it is crucial to rely on credible scientific evidence to separate fact from fiction and address the misconceptions that sustain the theory.

Peer-Reviewed Research and Publications

The bedrock of scientific integrity is peer-reviewed research, and it is through this rigorous process that the chemtrails conspiracy theory has been thoroughly examined and debunked. Peer-reviewed publications ensure that findings are scrutinized by experts in the field, validating the research methods and conclusions. In the case of contrails and chemtrails, numerous peer-reviewed studies have consistently shown that the trails left by aircraft are composed of water vapor and ice crystals, not harmful chemicals.

One notable study published in the journal Environmental Research Letters investigated the claims of chemtrails by analyzing the chemical composition of aircraft contrails and cirrus clouds. The researchers collected atmospheric data using high-altitude sampling and compared it with ground-based measurements. Their findings revealed that the chemical composition of contrails was consistent with the exhaust emissions from jet engines and the natural condensation process. The study found no evidence of unusual or harmful substances, thereby refuting the chemtrails claims.

Another key piece of research was conducted by the Carnegie Institution for Science, which focused on the potential environmental impact of aviation-induced cloud cover. The study, published in the Journal of Geophysical Research: Atmospheres, used satellite imagery and climate models to assess the effects of contrails on climate patterns. The results indicated that contrails can contribute to global warming by trapping infrared radiation, but there was no indication of intentional chemical spraying. The study emphasized that contrails are a byproduct of aviation, not a deliberate geoengineering effort.

In 2016, a comprehensive review of the chemtrails theory was published in the Journal of Atmospheric Chemistry and Physics. This review paper synthesized findings from multiple studies and expert opinions, concluding that there was no scientific basis for the chemtrails theory. The authors highlighted the importance of peer review in maintaining scientific credibility and noted that none of the studies supporting chemtrails claims had undergone this rigorous process. By contrast, all peer-reviewed research on contrails pointed to natural atmospheric processes as the cause of their formation and persistence.

The peer-review process itself is crucial in validating scientific research. When a study is submitted to a reputable journal, it undergoes scrutiny by other experts in the field who evaluate the methodology, data, and conclusions. This process helps to identify any biases, errors, or inconsistencies, ensuring that only high-quality research is published. The fact that chemtrails claims have not been supported by peer-reviewed studies underscores their lack of scientific validity.

Several other studies have contributed to debunking the chemtrails theory. For example, a study conducted by the University of Reading used high-altitude research aircraft to analyze contrail formation and its impact on weather patterns. Published in Nature Communications, the study found that contrails could influence cloud formation and local climate, but there was no evidence of chemical additives. The researchers concluded that the observed effects were consistent with known atmospheric science, not covert geoengineering.

In addition to specific studies, the broader scientific consensus reinforces the debunking of chemtrails. Organizations such as NASA, the National Oceanic and Atmospheric Administration (NOAA), and the World Meteorological Organization (WMO) have issued statements affirming that contrails are a natural consequence of jet aircraft emissions. These statements are based on extensive research and monitoring, providing a robust foundation for public understanding.

In conclusion, peer-reviewed research and publications play a pivotal role in debunking the chemtrails conspiracy theory. Through rigorous examination and validation by experts, these studies provide credible and reliable evidence that contrails are the result of natural atmospheric processes. The absence of peer-reviewed support for chemtrails claims further underscores their lack of scientific basis. As we continue to explore the world of chemtrails, it is essential to rely on credible scientific sources and peer-reviewed research to separate fact from fiction and promote an informed public discourse.

Role of Skeptics and Science Communicators

In the battle against misinformation and conspiracy theories, skeptics and science communicators play a crucial role. Their efforts to debunk the chemtrails theory and educate the public about contrails have been instrumental in promoting scientific literacy and critical thinking. These individuals and organizations work tirelessly to counteract the spread of false information, often facing significant challenges and pushback from conspiracy theorists.

Skeptics, often experts in fields such as atmospheric science, meteorology, and environmental research, dedicate their work to dispelling myths and providing accurate information. One notable figure is Mick West, a former video game programmer who has become a prominent debunker of conspiracy theories, including chemtrails. West runs the website Metabunk.org, where he meticulously examines and debunks various conspiracy claims. His approach involves detailed scientific explanations, debunking videos, and community discussions, all aimed at providing clear evidence-based counterarguments to chemtrail claims.

Another key figure is Dr. David Keith, a professor of applied physics and public policy at Harvard University. Dr. Keith is an expert in climate science and geoengineering and has been vocal about the misconceptions surrounding chemtrails. He emphasizes the importance of public understanding of legitimate scientific research in geoengineering, which is often conflated with the chemtrails theory. Through public lectures, interviews, and academic publications, Dr.

Keith educates the public on the differences between scientific research and conspiracy theories.

Science communicators also play a vital role in reaching a broader audience. These individuals use various media platforms to make complex scientific concepts accessible and engaging. Bill Nye, known as "The Science Guy," has addressed the chemtrails theory in several of his public appearances and videos. Nye uses his charismatic communication style to explain the science behind contrail formation and dispel the myths surrounding chemtrails. His efforts reach millions of viewers, making science more approachable and helping to counteract misinformation.

The strategies and methods used by skeptics and science communicators are diverse and adaptive. They include educational videos, detailed articles, social media engagement, public lectures, and interactive discussions. One effective approach is the use of visual aids and demonstrations to illustrate scientific principles. For example, videos showing the formation of contrails in controlled laboratory settings help demystify the process and provide tangible evidence against chemtrails claims. Infographics and animations that explain atmospheric conditions and contrail behavior are also valuable tools for educating the public.

Social media is a double-edged sword in the fight against misinformation. While it can spread false information quickly, it also offers a platform for science communicators to reach large audiences. Skeptics and educators use social media to share debunking articles, engage with followers, and participate in discussions. Platforms like Twitter and YouTube allow for real-time interaction, where experts can answer questions, clarify doubts, and provide evidence-based explanations.

Case studies of individuals who have changed their views on chemtrails through critical thinking and scientific education highlight the impact of these efforts. Testimonials from former believers who now understand the science behind contrails underscore the importance of outreach and education. These personal stories are powerful examples of how evidence-based reasoning and engagement with credible sources can shift perspectives and dispel conspiracy theories.

In conclusion, skeptics and science communicators play a vital role in debunking the chemtrails conspiracy theory and promoting scientific literacy. Through meticulous research, public engagement, and effective communication strategies, they provide clear, evidence-based counterarguments to chemtrail claims. Their work is essential in fostering critical thinking and helping the public navigate the complex landscape of misinformation. As we continue to explore the chemtrails narrative, the contributions of these individuals and organizations remain crucial for separating fact from fiction and promoting an informed and scientifically literate society.

The Importance of Critical Thinking and Skepticism

The persistence of the chemtrails conspiracy theory underscores the need for promoting critical thinking and skepticism. These cognitive tools are essential for evaluating conspiracy theories and distinguishing between credible information and misinformation. Critical thinking involves analyzing and evaluating evidence, questioning assumptions, and making reasoned judgments. Skepticism entails maintaining a questioning attitude and not accepting claims without sufficient evidence. Together, they form the foundation of scientific inquiry and rational decision-making.

One of the key aspects of critical thinking is the ability to recognize cognitive biases. Humans are prone to various biases that can influence how we interpret information. For example, confirmation bias leads individuals to seek out and give more weight to information that confirms their preexisting beliefs while ignoring or dismissing evidence that contradicts them. This bias is particularly prevalent in the context of conspiracy theories, where believers may selectively interpret data to fit the narrative of chemtrails. By being aware of such biases, individuals can strive to evaluate evidence more objectively and avoid falling into the trap of selective perception.

Another important element of critical thinking is the evaluation of sources. In the age of the internet, information is readily accessible, but not all sources are equally reliable. Evaluating the credibility of sources involves considering the expertise, background, and potential biases of the information provider. Peer-reviewed scientific journals, for instance, undergo rigorous review processes to ensure the quality and reliability of the research they publish. In contrast, many websites and social media platforms lack such scrutiny, making it easier for misinformation to spread. Encouraging individuals to prioritize credible sources and critically assess the validity of information is crucial in combating the spread of conspiracy theories.

Skepticism also plays a vital role in the scientific method. Science advances through questioning and testing hypotheses, relying on evidence and logical reasoning. This process involves being open to new information while requiring that claims be substantiated by empirical data. In the context of chemtrails, skepticism means questioning the extraordinary claims made by conspiracy theorists and demanding rigorous evidence to support them. The lack of concrete, reproducible evidence for chemtrails highlights the importance of maintaining a skeptical stance and not accepting claims at face value.

Educational initiatives aimed at fostering critical thinking and skepticism are essential for building a more scientifically literate society. Programs that teach these skills should be integrated into school curriculums and lifelong learning opportunities. Interactive workshops, science communication events, and public lectures can also engage the broader community and promote a culture of inquiry and evidence-based reasoning. By equipping individuals with the tools to critically evaluate information, we can reduce the influence of misinformation and promote a more informed public discourse.

Case studies of individuals who have changed their views on chemtrails through critical thinking highlight the transformative power of these cognitive tools. Testimonials from former believers who now understand the science behind contrails demonstrate that education and exposure to credible information can shift perspectives. These personal stories serve as powerful examples of how critical thinking and skepticism can counteract conspiracy theories and lead to a more accurate understanding of the world.

In conclusion, the importance of critical thinking and skepticism cannot be overstated in the effort to debunk the chemtrails conspiracy theory and other forms of misinformation. By promoting these cognitive tools, we empower individuals to make reasoned judgments, evaluate evidence objectively, and prioritize credible sources. Education and public awareness initiatives play a crucial role in fostering a culture of inquiry and evidence-based reasoning. As we continue to explore the chemtrails narrative, the promotion of critical thinking and skepticism remains essential for separating fact from fiction and building a more scientifically literate society.

Chapter 8: Psychological and Sociological Perspect

Why People Believe in Conspiracy Theories

Belief in conspiracy theories, such as chemtrails, is a complex phenomenon driven by a variety of psychological factors. One of the primary reasons people are drawn to these theories is the human propensity for pattern recognition. Our brains are wired to detect patterns and make connections, which was a crucial survival mechanism in our evolutionary past. However, this tendency can lead us to perceive meaningful patterns in random or unrelated events—a cognitive bias known as apophenia. When individuals see persistent contrails in the sky, their pattern-recognition mechanisms might interpret these as intentional and malicious actions rather than natural atmospheric phenomena.

Another cognitive bias that plays a significant role in conspiracy beliefs is confirmation bias. This bias leads people to seek out, interpret, and remember information that confirms their preexisting beliefs while disregarding evidence that contradicts them. For those inclined to distrust authority or believe in government malfeasance, any ambiguous or unexplained phenomenon can be readily interpreted as evidence supporting their worldview. In the case of chemtrails, believers might selectively focus on anomalous weather patterns or health issues and attribute them to chemical spraying, ignoring scientific explanations that refute these claims.

Uncertainty and fear are also powerful drivers of conspiracy thinking. In times of social, political, or environmental upheaval, individuals may experience heightened anxiety and a sense of loss of control. Conspiracy theories offer a way to make sense of a chaotic world by providing clear, albeit erroneous, explanations for complex events. Believing that there is a hidden hand orchestrating these events can be psychologically comforting, as it suggests that there is order and purpose behind apparent randomness. The chemtrails theory, with its narrative of covert government operations, provides a sense of understanding and control over inexplicable environmental and health phenomena.

Societal factors also contribute to the susceptibility to conspiracy theories. People who feel marginalized, disempowered, or distrustful of mainstream institutions are more likely to embrace alternative explanations that validate their experiences and perspectives. The rise of the internet and social media has further amplified this effect by creating echo chambers where like-minded individuals can reinforce each other's beliefs. Online communities dedicated to chemtrails and other conspiracy theories provide a sense of belonging and validation, making it easier for individuals to maintain and spread their beliefs.

The impact of societal and personal factors is evident in the diverse demographics of conspiracy believers. While some are motivated by political or ideological reasons, others might be driven by personal experiences or health concerns. For instance, individuals who have experienced unexplained health issues might be more inclined to believe in chemtrails as a potential cause, seeking explanations outside of conventional medical understanding.

In conclusion, belief in conspiracy theories like chemtrails is driven by a combination of psychological factors, including pattern recognition, confirmation bias, and the need for control and understanding in times of uncertainty. Societal influences, such as feelings of marginalization and the reinforcement of beliefs within online communities, also play a significant role. Recognizing these underlying factors is crucial for addressing and mitigating the spread of conspiracy theories. As we continue to explore the psychological and sociological perspectives of the chemtrails narrative, it becomes clear that fostering critical thinking and scientific literacy is essential for countering misinformation and promoting a more informed and rational public discourse.

The Psychological Appeal of Chemtrails

The chemtrails conspiracy theory holds a powerful psychological appeal for many believers. Several elements of the theory resonate emotionally and cognitively, providing a sense of meaning and coherence in a world that often feels chaotic and uncontrollable. Understanding these elements can help explain why the chemtrails narrative remains compelling despite substantial scientific debunking.

One key aspect of the psychological appeal is the emotional satisfaction derived from believing in chemtrails. For some individuals, the theory offers an explanation for various personal and societal issues that seem otherwise inexplicable. This sense of understanding can be deeply comforting, as it provides a narrative that ties together disparate events and phenomena. When faced with complex issues such as unusual weather patterns, environmental changes, or unexplained health problems, the idea of chemtrails offers a simple and clear explanation: a covert operation orchestrated by powerful entities. This narrative can be more palatable than the uncertainty and ambiguity inherent in scientific explanations.

The chemtrails theory also taps into a basic human need for control and agency. Believing in a conspiracy can give individuals a sense of empowerment, as it implies that they possess hidden knowledge that others do not. This perceived awareness sets them apart from the uninformed masses and positions them as enlightened and vigilant. The sense of having uncovered a secret can be intoxicating, providing a boost to one's self-esteem and reinforcing their belief in their own analytical abilities. In a world where individuals often feel powerless against larger social and political forces, the belief in chemtrails can restore a sense of control.

Moreover, the chemtrails community offers a strong sense of identity and belonging. Online forums, social media groups, and in-person meetups provide spaces where believers can connect with like-minded individuals, share information, and validate each other's experiences. These communities create an environment of mutual support and camaraderie, which can be especially appealing to those who feel isolated or marginalized in other aspects of their lives. The collective identity formed around the belief in chemtrails fosters a sense of solidarity and purpose, making it harder for individuals to abandon the theory even when faced with contradictory evidence.

The narrative of chemtrails also appeals to those who distrust authority and mainstream institutions. The theory often frames itself as a challenge to official narratives, positioning believers as skeptics who question the status quo. This rebellious stance can be particularly attractive to individuals who harbor a deep-seated mistrust of government, media, and scientific institutions. By aligning themselves with the chemtrails movement, they can feel part of a larger struggle against perceived corruption and deceit.

The cognitive appeal of the chemtrails theory is further enhanced by the sense of coherence and simplicity it provides. Conspiracy theories often thrive because they offer clear and definitive explanations for complex and multifaceted issues. The chemtrails narrative reduces the complexity of atmospheric science, environmental health, and geopolitics to a single, overarching plot. This simplicity can be more psychologically satisfying than the nuanced and sometimes inconclusive nature of scientific research. For believers, the theory ties up loose ends and presents a unified story that makes sense of the world.

In conclusion, the psychological appeal of the chemtrails conspiracy theory lies in its ability to provide emotional and cognitive satisfaction, a sense of control and identity, and a coherent narrative that challenges mainstream authority. These elements make the theory compelling and resistant to debunking efforts, as they resonate deeply with the human psyche. Understanding these psychological drivers is essential for addressing and mitigating the impact of conspiracy theories. As we continue to explore the psychological and sociological perspectives of the chemtrails narrative, it becomes clear that promoting critical thinking and fostering a sense of agency grounded in scientific understanding are crucial for countering misinformation and promoting rational discourse.

Sociological Factors Contributing to the Spread of the Theory

The spread of the chemtrails conspiracy theory is not solely driven by individual psychological factors but is also deeply rooted in sociological dynamics. Social networks, cultural contexts, and media landscapes play significant roles in facilitating the dissemination and entrenchment of the theory. Understanding these sociological factors is crucial for comprehending why the chemtrails narrative persists and how it gains traction among diverse populations.

Social networks, both online and offline, are critical in the spread of the chemtrails theory. Online platforms such as Facebook, YouTube, and Reddit provide spaces where individuals can share information, discuss their beliefs, and connect with like-minded individuals. These platforms create echo chambers, where users are exposed primarily to information that reinforces their preexisting beliefs. Algorithms that prioritize engagement often amplify sensationalist content, making conspiracy theories like chemtrails more visible and appealing. The sense of community and belonging that emerges in these online spaces can be powerful, encouraging individuals to adopt and spread the theory further.

Offline social networks also contribute to the dissemination of the chemtrails narrative. Friends, family members, and community groups can influence an individual's beliefs through direct interactions and shared experiences. In some cases, local events or meetings dedicated to discussing chemtrails and other conspiracy theories provide opportunities for believers to gather, exchange information, and strengthen their convictions. These face-to-face interactions can be particularly impactful, as they foster a sense of solidarity and collective identity.

Cultural factors also play a significant role in shaping public perception of the chemtrails theory. Societal attitudes toward authority, science, and technology can influence the likelihood of individuals embracing conspiracy theories. In cultures with a strong tradition of skepticism toward government and scientific institutions, conspiracy theories may find a more receptive audience. Historical events that have eroded public trust, such as governmental scandals or controversial scientific developments, can create a fertile ground for conspiracy thinking. The chemtrails theory taps into these cultural undercurrents, framing itself as a challenge to corrupt and deceptive authorities.

The media landscape is another crucial factor in the spread of the chemtrails theory. Sensationalist reporting and the proliferation of alternative media outlets have contributed to the visibility and appeal of conspiracy narratives. Mainstream media sometimes cover conspiracy theories in a way that grants them undue credibility, either by presenting them as legitimate controversies or by failing to adequately debunk them. Alternative media, which often positions itself as a counterpoint to mainstream narratives, provides a platform for conspiracy theories to thrive. These outlets may frame the chemtrails theory as an "expose" of hidden truths, attracting audiences who are already inclined to question official accounts.

The influence of social identity and group dynamics is also significant. Believing in the chemtrails theory can become a marker of group membership, differentiating believers from non-believers. This social identity can be reinforced through rituals, such as sharing evidence, attending meetings, or participating in online discussions. Group dynamics, including peer pressure and the desire for social cohesion, can further entrench these beliefs. Individuals who are part of a community of chemtrails believers may feel a sense of loyalty and commitment to the group, making it difficult to abandon the theory even in the face of contradicting evidence.

In conclusion, the spread of the chemtrails conspiracy theory is facilitated by a complex interplay of sociological factors, including social networks, cultural attitudes, media dynamics, and group identity. These elements create an environment where the theory can flourish, providing social validation and reinforcement for believers. Understanding these sociological dimensions is essential for developing strategies to address and mitigate the impact of conspiracy theories. As we continue to explore the chemtrails narrative, it is clear that promoting media literacy, fostering critical thinking, and building trust in credible sources are crucial steps in countering misinformation and promoting a more informed public discourse.

The Impact on Individuals and Communities

Believing in the chemtrails conspiracy theory can have significant psychological and social impacts on individuals and communities.

These impacts manifest in various aspects of life, from mental health to interpersonal relationships, and extend to broader societal implications.

At an individual level, adherence to the chemtrails theory can lead to increased anxiety and fear. The belief that one is constantly being subjected to harmful chemicals sprayed from the sky can create a pervasive sense of threat and paranoia. This heightened state of alertness and suspicion can contribute to chronic stress, which negatively affects both mental and physical health. Individuals may experience symptoms such as insomnia, headaches, and digestive issues, which are exacerbated by the constant worry about perceived dangers.

The impact on mental health can also extend to feelings of isolation and alienation. Believers in the chemtrails theory often find themselves at odds with friends, family members, and colleagues who do not share their views. This divergence in beliefs can strain relationships, leading to conflicts and social withdrawal. Individuals may feel misunderstood and marginalized, reinforcing their sense of being part of a persecuted minority. This social isolation can further entrench their beliefs, as they seek validation and support within conspiracy theory communities.

Within these communities, the sense of belonging and mutual support can be both comforting and reinforcing. Online forums, social media groups, and local meetups provide spaces where believers can share their experiences, exchange information, and support one another. While these communities offer a sense of solidarity, they can also create echo chambers that amplify and reinforce conspiracy beliefs. The constant validation and affirmation within these groups make it difficult for individuals to question or abandon their beliefs, even when confronted with contradicting evidence.

The chemtrails conspiracy theory can also have broader social implications, particularly in terms of trust in institutions. Believers often harbor deep-seated mistrust of government, scientific institutions, and the media. This erosion of trust can undermine public confidence in these institutions, making it more challenging for them to effectively communicate and implement policies, especially in areas such as public health and environmental protection. The spread of conspiracy theories can also polarize communities, creating divisions based on differing beliefs and worldviews.

In extreme cases, belief in the chemtrails theory can lead to real-world actions that have negative consequences. Some individuals may engage in activism based on their beliefs, such as protesting against perceived chemtrail spraying or attempting to interfere with aviation operations. While most actions are non-violent, they can create tensions and disrupt public order. Additionally, individuals may make health and lifestyle decisions based on their conspiracy beliefs, such as avoiding certain areas, purchasing expensive "detox" products, or rejecting medical advice, which can have detrimental effects on their well-being.

The broader societal implications of widespread conspiracy beliefs extend to the realm of democracy and civic engagement. When large segments of the population subscribe to conspiracy theories, it can erode the foundations of informed and rational discourse. Public debates and policy discussions can become muddled by misinformation, making it difficult to reach consensus on important issues. The distrust in institutions and experts fostered by conspiracy theories can also lead to decreased participation in democratic processes, as individuals feel disillusioned and powerless to effect change.

In conclusion, belief in the chemtrails conspiracy theory can significantly impact individuals and communities, affecting mental health, relationships, trust in institutions, and civic engagement.

These impacts highlight the importance of addressing and mitigating the spread of conspiracy theories through education, critical thinking, and community support. As we continue to explore the psychological and sociological perspectives of the chemtrails narrative, it becomes clear that fostering a culture of skepticism, scientific literacy, and informed public discourse is essential for countering misinformation and promoting a healthy, cohesive society.

Strategies for Addressing and Mitigating Belief in Chemtrails

Addressing and mitigating the belief in the chemtrails conspiracy theory requires a multifaceted approach that combines education, critical thinking, media literacy, and community support. These strategies aim to provide individuals with the tools and knowledge to evaluate information critically, reduce the appeal of conspiracy theories, and foster trust in credible sources.

Education is the cornerstone of any strategy to counteract conspiracy theories. Integrating critical thinking and scientific literacy into school curriculums from an early age can equip individuals with the skills to question and assess information effectively. Educational programs should emphasize the importance of evidence-based reasoning, the scientific method, and the evaluation of sources. By fostering an understanding of how scientific knowledge is generated and validated, we can help individuals discern between credible information and misinformation.

Media literacy is another crucial component in addressing conspiracy beliefs. In the digital age, where information is readily accessible but not always reliable, teaching individuals how to navigate the media landscape is essential. Media literacy programs should focus on developing the skills to critically evaluate online content, recognize biases, and identify credible sources. Encouraging individuals to question the authenticity and accuracy of information they encounter can help reduce the spread of conspiracy theories. Resources such as fact-checking websites and guidelines for assessing the credibility of information should be widely promoted.

Community support and outreach are also vital in addressing the social aspects of conspiracy beliefs. Building trust and open communication within communities can create an environment where individuals feel comfortable discussing their beliefs and concerns. Community leaders, educators, and healthcare professionals can play a significant role in fostering dialogue and providing accurate information. Support groups and community events focused on science and critical thinking can offer a positive alternative to conspiracy theory communities, providing social validation and a sense of belonging without relying on misinformation.

Interventions targeting individuals heavily invested in the chemtrails theory must be approached with sensitivity and respect. Confrontational or dismissive tactics are likely to be counterproductive, as they can reinforce individuals' defensive attitudes and strengthen their commitment to their beliefs. Instead, engaging in respectful and empathetic dialogue can create opportunities for individuals to reflect on their beliefs and consider alternative perspectives. Motivational interviewing techniques, which focus on exploring and resolving ambivalence, can be effective in encouraging individuals to question their beliefs and seek out credible information.

Promoting critical thinking and scientific literacy more broadly across society is essential for reducing the appeal of conspiracy theories. Public awareness campaigns can highlight the importance of skepticism and evidence-based reasoning in everyday life. Collaborations between scientists, educators, and media professionals can help disseminate accurate information and counteract misinformation. Platforms for public science communication, such as public lectures, science fairs, and media appearances by experts, can engage the broader community and promote a culture of inquiry and rational thinking.

In conclusion, addressing and mitigating belief in the chemtrails conspiracy theory requires a comprehensive and multifaceted approach. Education, media literacy, community support, and respectful dialogue are all essential components of this strategy. By equipping individuals with the skills to critically evaluate information, fostering open communication, and promoting scientific literacy, we can reduce the appeal of conspiracy theories and build a more informed and rational society. Understanding and addressing the psychological and sociological factors that drive belief in chemtrails is crucial for developing effective interventions and promoting a healthier public discourse.

Chapter 9: Media and Pop Culture Influence

Representation in Mainstream Media

The representation of chemtrails in mainstream media has played a significant role in shaping public perception and spreading the conspiracy theory. News outlets, television shows, and movies have all contributed to the visibility of chemtrails, often with varying degrees of sensationalism and credibility.

In news reporting, chemtrails have occasionally been presented as a legitimate controversy, giving the theory an undue sense of credibility. Sensationalist headlines and dramatic visuals can attract viewers and readers, but they also risk amplifying misinformation. For instance, news segments featuring dramatic footage of crisscrossing trails in the sky, accompanied by ominous music and speculative commentary, can create a sense of alarm and legitimacy around the theory. While responsible journalism aims to present balanced and well-researched information, the pressure to capture audience attention can sometimes lead to sensationalist coverage that fuels conspiracy beliefs.

A notable example of media coverage influencing public perception occurred in the early 2000s when several local news stations in the United States aired segments on chemtrails. These reports often featured interviews with concerned citizens and "experts" who claimed to have evidence of chemical spraying. The absence of rigorous scientific rebuttal in these segments allowed the theory to gain traction and spread more widely. The media's role in this context was not to deliberately misinform but to provide coverage of a topic that had garnered public interest. However, the lack of critical examination and reliance on anecdotal evidence contributed to the theory's persistence.

Television shows and movies have also played a role in popularizing the chemtrails theory. Sci-fi and thriller genres, in particular, have featured plotlines involving government conspiracies and covert operations, which can resonate with audiences already inclined to distrust authority. Shows like "The X-Files" have depicted scenarios where the government engages in secretive and sinister activities, reinforcing the idea that such operations are plausible. While these shows are intended for entertainment, they can blur the line between fiction and reality for some viewers, making it easier for conspiracy theories like chemtrails to take root.

Movies have similarly contributed to the visibility of chemtrails. Films that explore themes of governmental control and environmental manipulation can inadvertently lend credence to the theory. For example, the 1998 film "The Siege" depicted a scenario where the U.S. government imposes martial law, feeding into fears of governmental overreach. Although the film does not specifically address chemtrails, its portrayal of a secretive and powerful government can reinforce the broader narrative of conspiracy theories.

The balance between responsible journalism and sensationalism is a critical issue in the context of chemtrails. Media outlets have a responsibility to provide accurate and well-researched information, but they also operate in a competitive environment where capturing audience attention is paramount. Striking this balance is challenging, as sensationalist reporting can generate higher ratings and more clicks but at the expense of spreading misinformation.

In conclusion, the representation of chemtrails in mainstream media has significantly influenced public perception and the spread of the conspiracy theory. Sensationalist reporting, dramatic portrayals in television shows and movies, and the challenges of balancing responsible journalism with audience engagement have all contributed to the visibility and persistence of the chemtrails narrative. Understanding the role of media in shaping beliefs is crucial for developing strategies to counteract misinformation and promote accurate, evidence-based information. As we continue to explore the influence of media and pop culture on the chemtrails theory, it is essential to foster critical media literacy and support responsible journalism that prioritizes accuracy over sensationalism.

Influence of Documentaries and Independent Films

Documentaries and independent films have played a significant role in popularizing and spreading the chemtrails conspiracy theory. These visual media formats offer a powerful way to communicate complex ideas and narratives, often with an emotional and dramatic impact that resonates deeply with audiences. By presenting chemtrails as a hidden truth or a suppressed reality, these films can attract viewers who are already skeptical of mainstream narratives and authorities.

One of the key documentaries that has influenced public perception of chemtrails is "What in the World Are They Spraying?" Released in 2010, this film explores the chemtrails theory in depth, featuring interviews with self-proclaimed experts, activists, and concerned citizens. The documentary presents anecdotal evidence and speculative interpretations, framing chemtrails as a deliberate and harmful operation carried out by governments and corporations.

The film's dramatic tone, coupled with its compelling narrative, has made it a cornerstone in the chemtrails community, reinforcing and spreading the conspiracy theory.

Another influential documentary is "Why in the World Are They Spraying?" a follow-up to the 2010 film. This sequel delves further into the alleged motivations behind chemtrails, such as weather modification, population control, and corporate profit. It builds on the foundation laid by the first film, providing more interviews, testimonies, and visual evidence to support the theory. The impact of these documentaries is evident in the way they are widely shared and discussed within conspiracy theory forums and social media groups, reaching a broad audience and legitimizing the chemtrails narrative.

Independent films, often produced with lower budgets and fewer resources than mainstream productions, can also have a significant impact on public perception. These films are frequently shared through alternative media channels, bypassing traditional distribution networks and reaching audiences directly. The grassroots nature of these productions can lend an air of authenticity and credibility to the messages they convey, appealing to viewers who are distrustful of mainstream media. Independent filmmakers often position themselves as truth-seekers, challenging official narratives and exposing hidden realities, which resonates with conspiracy theorists.

The influence of visual media, particularly documentaries, on belief in chemtrails is substantial because of the way it engages viewers emotionally and cognitively. Visual storytelling allows for the presentation of dramatic and compelling evidence, such as aerial footage of crisscrossing trails and interviews with passionate advocates. This format can create a sense of urgency and importance, making the viewer feel part of a larger movement to uncover the truth. The emotional impact of these films can be profound, fostering a sense of solidarity and shared mission among viewers.

Case studies of prominent films illustrate their impact on the chemtrails narrative. For example, the "Geoengineering Watch" series, produced by Dane Wigington, combines professional-looking visuals with scientific-sounding commentary to argue that chemtrails are part of a vast geoengineering effort. These films are widely circulated within conspiracy theory communities, often serving as entry points for new believers. The credibility lent by the visual medium, along with the authoritative tone of the narration, can make it challenging for viewers to critically evaluate the claims presented.

In conclusion, documentaries and independent films have significantly influenced the spread and persistence of the chemtrails conspiracy theory. By leveraging the emotional and cognitive power of visual storytelling, these media formats can communicate complex and speculative ideas in a compelling manner. The impact of these films is evident in the way they are shared and discussed within conspiracy theory communities, reinforcing and legitimizing the chemtrails narrative. As we continue to explore the influence of media and pop culture on the chemtrails theory, it is essential to promote media literacy and critical thinking to help viewers critically evaluate the information presented in visual media.

Role of Social Media Platforms

Social media platforms have been instrumental in the proliferation and persistence of the chemtrails conspiracy theory. Platforms like Facebook, YouTube, and Twitter provide fertile ground for the rapid dissemination of chemtrails content, creating echo chambers where misinformation can flourish and spread unchallenged. The dynamics of these platforms, driven by algorithms designed to maximize engagement, often amplify sensationalist and conspiratorial content, making it highly visible and accessible to a wide audience.

Facebook, with its vast user base and extensive reach, has become a central hub for chemtrails believers. Groups and pages dedicated to the theory serve as gathering places where individuals can share their experiences, post photos and videos of supposed chemtrails, and discuss their interpretations of these observations. These communities offer a sense of validation and belonging, reinforcing the belief in chemtrails and making it more resistant to outside challenge. The platform's algorithm, which prioritizes content that generates high levels of engagement, can inadvertently promote conspiracy posts that elicit strong emotional reactions, further entrenching the narrative.

YouTube also plays a significant role in spreading chemtrails content through videos and documentaries. The platform's recommendation algorithm, designed to keep users engaged for longer periods, often suggests related videos that align with the viewer's interests. This can lead users down a rabbit hole of conspiracy content, as one video on chemtrails leads to another, each reinforcing and expanding on the same narrative. Popular YouTube channels dedicated to conspiracy theories, including chemtrails, have amassed large followings and millions of views, demonstrating the platform's power to shape public perception.

Twitter, with its real-time information sharing and trending hashtags, enables the rapid spread of chemtrails content across the globe. Hashtags like #chemtrails and #geoengineering allow users to join larger conversations, connect with others who share their beliefs, and amplify their messages. The platform's brevity and immediacy make it an effective tool for spreading concise, impactful claims, often accompanied by striking images or videos. Influencers and activists on Twitter can quickly mobilize their followers around chemtrails content, contributing to the theory's visibility and reach.

The role of influencers and content creators cannot be overlooked. These individuals, who often position themselves as independent investigators or truth-seekers, produce content that appeals to conspiracy-minded audiences. They leverage their platforms to present themselves as authorities on the subject, using a mix of scientific-sounding language and emotional appeals to convince viewers of the validity of the chemtrails theory. Their influence extends beyond social media, as they are often featured in documentaries, interviews, and public events, further legitimizing the narrative.

Regulating misinformation on social media presents significant challenges. While platforms have implemented measures to combat false information, such as fact-checking and content moderation, these efforts are often met with resistance from conspiracy theorists who view them as censorship. Balancing the need to address misinformation with the protection of free speech rights is a complex issue that requires careful consideration. Collaborative efforts between social media companies, fact-checking organizations, and public institutions are essential for developing effective strategies to mitigate the spread of conspiracy theories like chemtrails.

In conclusion, social media platforms play a crucial role in the dissemination and entrenchment of the chemtrails conspiracy theory. The algorithms and dynamics of these platforms, driven by engagement, often amplify sensationalist and conspiratorial content, making it highly visible and accessible. Influencers and content creators leverage their platforms to spread and legitimize the narrative, while challenges in regulating misinformation complicate efforts to address the issue. Understanding the role of social media in shaping beliefs is essential for developing strategies to counteract misinformation and promote accurate, evidence-based information. As we continue to explore the influence of media and pop culture on the chemtrails theory, fostering critical media literacy and supporting responsible content moderation remain key priorities.

Pop Culture References and Their Impact

Pop culture references to chemtrails in music, literature, and other forms of entertainment have played a significant role in both legitimizing and popularizing the conspiracy theory. These references often embed the chemtrails narrative in engaging and relatable contexts, making the idea more accessible and compelling to a broader audience. By exploring how chemtrails are depicted in various cultural products, we can better understand their impact on public perception and belief.

Music has been a powerful medium for expressing and disseminating conspiracy theories, including chemtrails. Artists across different genres have incorporated chemtrails into their lyrics, using their platform to highlight the theory and question official narratives. For instance, rapper B.o.B released a track titled "Flatline" in which he mentions chemtrails, linking them to broader themes of governmental distrust and manipulation. The song's popularity helped bring the chemtrails theory to the attention of a wider audience, particularly among younger listeners who may not have been previously aware of the narrative. The integration of chemtrails into music not only raises awareness but also lends an air of cultural legitimacy to the theory.

Literature has also played a role in embedding chemtrails into the cultural consciousness. Both fiction and non-fiction books have explored the theory, often framing it within larger narratives of environmental and political intrigue. In fiction, chemtrails might appear as a plot device in thrillers and dystopian novels, where secretive government programs and environmental manipulation serve as key elements of the story. These fictional depictions can make the idea of chemtrails more plausible by weaving them into coherent and engaging narratives. Non-fiction books, particularly those written by conspiracy theorists, delve into alleged evidence and testimonies, presenting chemtrails as a real and pressing issue. These works often appeal to readers who are already skeptical of mainstream explanations and are seeking alternative perspectives.

Television shows and movies have further contributed to the visibility and normalization of the chemtrails narrative. Popular series like "The X-Files," which frequently delved into government conspiracies and unexplained phenomena, have depicted scenarios that align with the chemtrails theory. Although these portrayals are fictional, they can influence viewers' perceptions by making the idea of chemtrails seem more plausible. The suspense and intrigue inherent in these narratives capture the audience's imagination, blurring the line between entertainment and reality.

Pop culture references extend beyond traditional media to include online content and memes. Social media platforms are rife with memes that reference chemtrails, often using humor and satire to comment on the theory. While some memes aim to debunk or mock the idea, others reinforce it by presenting chemtrails as an established fact. The viral nature of memes allows them to reach a vast audience quickly, embedding the chemtrails narrative in everyday online discourse. This normalization can desensitize the public to the more outrageous aspects of the theory, making it easier for individuals to accept chemtrails as a plausible explanation for observed phenomena.

The impact of these pop culture references is significant. By embedding chemtrails in music, literature, television, and online content, the theory becomes a familiar and relatable part of the cultural landscape. This familiarity can lend the theory an air of legitimacy, as individuals are more likely to believe in ideas that are repeatedly presented in engaging and authoritative contexts. Moreover, the emotional and cognitive engagement elicited by these cultural products can make the chemtrails narrative more compelling and memorable, reinforcing belief and spreading the theory to new audiences.

In conclusion, pop culture references to chemtrails in music, literature, television, and online content play a crucial role in legitimizing and popularizing the conspiracy theory. By embedding the narrative in engaging and relatable contexts, these references make the idea more accessible and compelling to a broader audience. Understanding the impact of pop culture on the spread of the chemtrails theory highlights the need for media literacy and critical thinking to navigate and evaluate the cultural products we consume. As we continue to explore the influence of media and pop culture on the chemtrails narrative, fostering a discerning and informed public remains essential.

The Role of Alternative Media and Citizen Journalism

The rise of alternative media and citizen journalism has significantly contributed to the spread and entrenchment of the chemtrails conspiracy theory. These non-traditional sources of information often position themselves as truth-seekers and watchdogs, challenging mainstream narratives and providing a platform for voices that feel marginalized or ignored by conventional media outlets. While these platforms can play an important role in promoting diverse perspectives, they also have the potential to amplify misinformation and conspiracy theories.

Alternative media outlets, such as independent news websites, blogs, and online forums, have become central hubs for the chemtrails community. These platforms often operate outside the constraints of traditional journalistic standards, allowing for a greater degree of editorial freedom. This freedom can be both a strength and a weakness; while it enables the exploration of unconventional viewpoints, it also means that claims can be published without rigorous fact-checking or peer review. As a result, alternative media can become echo chambers where conspiracy theories like chemtrails are disseminated and reinforced without critical scrutiny.

Prominent figures in alternative media have played key roles in promoting the chemtrails narrative. For example, websites like Infowars, run by Alex Jones, have featured extensive coverage of chemtrails, presenting them as part of a larger web of governmental and corporate conspiracies. Jones and other influential voices in alternative media often use sensationalist language and emotional appeals to engage their audience, framing chemtrails as an urgent and existential threat. This approach can create a sense of immediacy and legitimacy around the theory, attracting viewers who are predisposed to distrust mainstream institutions.

Citizen journalism, facilitated by the widespread availability of digital technology, has also played a significant role in the spread of the chemtrails theory. Individuals armed with smartphones and internet access can document and share their observations of the sky, often interpreting persistent contrails as evidence of chemical spraying. Platforms like YouTube, Facebook, and Twitter provide spaces for these citizen journalists to publish their content and reach a broad audience. The personal and grassroots nature of this content can make it particularly persuasive, as it appears to come from ordinary people rather than distant or impersonal institutions.

The credibility and influence of alternative media and citizen journalism are shaped by their perceived authenticity and alignment with the values of their audience. For individuals who feel alienated by mainstream media, these alternative sources can offer a sense of community and validation. The narratives presented by alternative media often resonate with the lived experiences and suspicions of their audience, making them more likely to be accepted as truth.

This dynamic creates a feedback loop where beliefs are reinforced and skepticism toward mainstream sources is heightened.

However, the potential for alternative media to inform and empower the public is tempered by the risk of spreading misinformation. The lack of editorial oversight and the emphasis on sensationalism can lead to the amplification of unfounded claims and conspiracy theories. Efforts to counteract this misinformation must balance the need to respect diverse viewpoints with the imperative to promote accurate and evidence-based information. Fact-checking initiatives and collaborations between mainstream and alternative media can help bridge this gap, providing a more balanced and informed discourse.

In conclusion, alternative media and citizen journalism have played a significant role in the spread and entrenchment of the chemtrails conspiracy theory. While these platforms offer valuable perspectives and a sense of community for their audiences, they also have the potential to amplify misinformation. Understanding the role of alternative media in shaping beliefs is crucial for developing strategies to address and mitigate the impact of conspiracy theories. Promoting media literacy, critical thinking, and collaboration between diverse media sources can help create a more informed and rational public discourse. As we continue to explore the influence of media and pop culture on the chemtrails narrative, fostering a discerning and well-informed audience remains essential.

Chapter 10: Case Studies and Real-World Impact

Notable Chemtrails Incidents and Investigations

The chemtrails conspiracy theory has been fueled by several notable incidents that proponents interpret as evidence of chemical spraying. These events have been the subject of various investigations conducted by scientists, government agencies, and independent researchers, each seeking to uncover the truth behind the mysterious trails observed in the sky. By examining these incidents and the subsequent investigations, we can better understand how they have influenced public belief and the persistence of the chemtrails narrative.

One of the most widely cited incidents occurred in 2000 when residents of Espanola, Ontario, reported persistent trails in the sky and a subsequent increase in respiratory illnesses and flu-like symptoms. Concerned citizens petitioned the Canadian government to investigate the phenomenon, suspecting that the trails were the result of chemical spraying. The government conducted a thorough investigation, including air quality testing and analysis of environmental samples. The results indicated that the observed trails were consistent with contrails, which are composed of water vapor and ice crystals, and found no evidence of harmful substances. Despite these findings, the incident continues to be referenced by chemtrails proponents as proof of covert operations.

Another significant case is the 2007 Los Angeles incident, where residents reported unusual patterns in the sky, accompanied by a purported increase in respiratory issues and environmental changes. This event prompted local media coverage and public outcry, leading to an investigation by environmental agencies. The Los Angeles County Department of Public Health conducted air quality tests and reviewed health data, concluding that there were no unusual or harmful substances detected in the air. The patterns observed were attributed to atmospheric conditions conducive to contrail formation and persistence. However, the incident added fuel to the chemtrails narrative, with believers citing it as evidence of chemical spraying.

In the United Kingdom, the 2011 Cornwall incident garnered significant attention. Residents reported seeing persistent trails in the sky over several days, which they believed were linked to an increase in respiratory illnesses and environmental changes. The UK government responded by conducting a comprehensive investigation, involving the Meteorological Office and the Department for Environment, Food & Rural Affairs (DEFRA). The investigation confirmed that the trails were contrails resulting from regular commercial flights and that the atmospheric conditions at the time were ideal for their formation. No harmful substances were found in the air or environmental samples. Despite these findings, the incident remains a focal point for chemtrails advocates in the UK.

In addition to these high-profile cases, numerous smaller incidents worldwide have contributed to the spread of the chemtrails theory. Each incident typically follows a similar pattern: residents observe persistent trails in the sky, report health issues or environmental changes, and call for investigations. While the investigations consistently find no evidence of chemical spraying and attribute the phenomena to natural atmospheric processes, the persistence of these incidents keeps the theory alive.

The influence of these incidents on public belief is substantial. Each case adds to the narrative that chemtrails are a widespread and ongoing operation, supported by anecdotal evidence and media coverage. The investigations, despite debunking the theory, often receive less attention than the initial reports, allowing the belief in chemtrails to persist. The tendency of believers to view the investigations as part of a cover-up further entrenches the narrative, making it resistant to debunking efforts.

In conclusion, notable chemtrails incidents and investigations play a crucial role in shaping public belief and the persistence of the chemtrails conspiracy theory. Despite consistent findings from scientific and governmental investigations that attribute the observed phenomena to natural atmospheric processes, the narrative of chemical spraying remains compelling for many. Understanding the impact of these incidents is essential for addressing and mitigating the spread of the chemtrails theory, emphasizing the need for transparent communication, public education, and critical thinking.

Legal and Political Responses

The chemtrails conspiracy theory has not only captivated public imagination but has also prompted legal and political actions. These responses, ranging from lawsuits to legislative measures and government inquiries, highlight the significant impact that the theory has had on public discourse and institutional trust. Examining these legal and political actions provides insight into how the theory has been addressed at various levels and the broader implications of these efforts.

One of the most notable legal actions related to chemtrails occurred in 2002 when a group of citizens from Santa Cruz, California, filed a lawsuit against the federal government. The plaintiffs claimed that the government was conducting clandestine weather modification operations through chemtrails, leading to environmental and health issues. They sought an injunction to halt the alleged spraying and demanded full disclosure of the activities. The lawsuit was ultimately dismissed due to a lack of evidence, with the court ruling that the plaintiffs had not provided credible scientific data to support their claims. This case underscored the challenges of proving conspiracy theories in a legal context, where the burden of proof requires substantial and credible evidence.

In a similar vein, the state of Shasta County, California, witnessed significant public outcry over chemtrails in 2014. During a Board of Supervisors meeting, concerned citizens presented their claims about chemtrails and the supposed environmental and health impacts. The board responded by agreeing to formally investigate the claims, which led to a series of hearings and expert testimonies. Despite extensive investigations and input from environmental scientists, the findings echoed previous investigations: there was no credible evidence of chemical spraying, and the observed phenomena were attributed to contrails formed under specific atmospheric conditions. The political response in this case demonstrated a commitment to addressing public concerns, but also highlighted the difficulty of dispelling deeply held beliefs.

Legislative measures have also been introduced in response to the chemtrails theory. For instance, in 2010, an Arizona state legislator proposed a bill to ban weather modification programs and require transparency in atmospheric activities. The bill, known as the "Weather Modification and Geoengineering Control Act," aimed to address public fears about chemtrails and ensure that any atmospheric interventions were subject to public oversight. Although the bill did not pass, it reflected the growing influence of chemtrails beliefs on political agendas and the efforts of some lawmakers to respond to their constituents' concerns.

Government inquiries have played a crucial role in addressing chemtrails claims. In the United Kingdom, the House of Commons Science and Technology Committee conducted an inquiry into geoengineering in 2009, which included a review of the chemtrails theory. The committee's report concluded that there was no evidence to support the existence of chemtrails and that geoengineering research should be conducted transparently and with public engagement. Similarly, the German Bundestag's Committee on Environment, Nature Conservation, and Nuclear Safety examined chemtrails in 2013, concluding that the theory was not supported by scientific evidence. These inquiries have helped to clarify official positions on chemtrails and provide authoritative rebuttals to the claims.

The impact of legal and political responses to the chemtrails theory extends beyond the immediate outcomes of investigations and legislation. These actions shape public perception and trust in institutions. For believers in chemtrails, the dismissal of lawsuits and the conclusions of government inquiries are often interpreted as further evidence of a cover-up, reinforcing their skepticism. Conversely, for skeptics and those seeking accurate information, these responses provide reassurance and a basis for debunking the theory.

In conclusion, legal and political responses to the chemtrails conspiracy theory highlight the significant influence the theory has had on public discourse and institutional trust. While lawsuits, legislative measures, and government inquiries have consistently found no evidence to support the theory, they also underscore the challenges of addressing deeply held beliefs through legal and political channels. Understanding these responses and their impact is essential for developing strategies to address the spread of conspiracy theories and promote public trust in credible sources and institutions. As we continue to explore the real-world impact of the chemtrails narrative, fostering transparent communication and evidence-based decision-making remains crucial.

Impact on Public Health and the Environment

The chemtrails conspiracy theory posits that the chemicals allegedly being sprayed from aircraft have a significant impact on public health and the environment. Proponents of the theory argue that chemtrails are responsible for a range of health issues, including respiratory problems, neurological disorders, and weakened immune systems. They also claim that these chemicals cause environmental damage, affecting soil quality, water sources, and ecosystems. To address these concerns, various studies and reports have been conducted to evaluate the perceived and actual impact of chemtrails on public health and the environment.

One of the primary health concerns cited by chemtrails believers is the increase in respiratory issues, such as asthma and bronchitis, which they attribute to exposure to chemicals dispersed in the atmosphere. Anecdotal reports of individuals experiencing persistent coughs, shortness of breath, and other respiratory symptoms are frequently shared within the chemtrails community. These accounts often coincide with observations of persistent contrails in the sky, leading to the conclusion that chemtrails are the cause. However, extensive air quality monitoring and epidemiological studies have found no link between contrails and adverse health effects. Research conducted by environmental agencies, such as the Environmental Protection Agency (EPA) and the World Health Organization (WHO), has consistently shown that the levels of chemicals like aluminum and barium in the air are within safe limits and not associated with aircraft emissions.

Neurological disorders are another area of concern for chemtrails proponents. They claim that exposure to the alleged chemicals in chemtrails can lead to cognitive decline, memory loss, and other neurological issues. These claims are often based on anecdotal evidence and speculative interpretations of scientific studies. However, peer-reviewed research has not supported these assertions. Studies examining the potential neurotoxic effects of aluminum and other substances cited by chemtrails believers have found no evidence that exposure from aircraft contrails poses a risk to neurological health. The observed symptoms can typically be attributed to other environmental and lifestyle factors, such as pollution, stress, and aging.

The environmental impact of chemtrails is also a significant concern for believers. They argue that the chemicals allegedly dispersed from aircraft can contaminate soil, water sources, and ecosystems, leading to negative consequences for agriculture, wildlife, and natural habitats. Claims of soil acidification, water contamination, and damage to crops and vegetation are commonly cited within the chemtrails narrative. To address these concerns, scientists and environmental researchers have conducted numerous studies on the environmental impact of contrails and aircraft emissions. The findings consistently indicate that contrails are composed of water vapor and ice crystals, with no evidence of harmful chemicals being dispersed. The environmental changes observed by chemtrails proponents can often be explained by other factors, such as industrial pollution, agricultural practices, and climate change.

Public health and environmental agencies have taken steps to address the concerns raised by chemtrails proponents. Outreach and educational campaigns aim to provide accurate information about contrails and their effects, countering the misinformation spread by the chemtrails theory. These efforts emphasize the importance of relying on credible scientific research and environmental monitoring to understand the true impact of atmospheric phenomena.

In conclusion, the perceived impact of chemtrails on public health and the environment is a central component of the conspiracy theory. However, extensive research and monitoring conducted by environmental and public health agencies have consistently found no evidence to support these claims. The health and environmental issues attributed to chemtrails can typically be explained by other factors, underscoring the importance of relying on credible scientific sources for information. As we continue to explore the real-world impact of the chemtrails narrative, it is essential to promote public education, critical thinking, and trust in scientific research to address and mitigate the spread of misinformation.

Community Reactions and Activism

The chemtrails conspiracy theory has not only sparked public interest and controversy but also led to significant community reactions and activism. Across the globe, individuals and groups have organized efforts to raise awareness, demand action, and push for investigations into chemtrails. These grassroots movements have utilized a variety of methods and strategies to spread their message and influence public opinion, highlighting the social dynamics and impact of chemtrails activism within communities.

One of the key aspects of community reactions to the chemtrails theory is the formation of local and national advocacy groups. These organizations, such as "Skywatch" in the United States and "The Chemtrails Project" in the United Kingdom, have played a central role in mobilizing citizens, organizing events, and distributing information. These groups often host public meetings, workshops, and rallies to discuss the theory and share their concerns with the broader community. By providing a platform for like-minded individuals to connect and collaborate, these advocacy groups foster a sense of solidarity and collective action.

Social media has been a crucial tool for chemtrails activists, allowing them to reach a wide audience and coordinate their efforts effectively. Platforms like Facebook, Twitter, and Instagram are used to share photos and videos of alleged chemtrails, disseminate articles and documentaries, and promote events and campaigns. Online petitions, such as those on Change.org, have garnered thousands of signatures, calling for government investigations and action against chemtrails. The viral nature of social media enables activists to amplify their message quickly and mobilize supporters on a global scale.

One prominent example of chemtrails activism is the "Global March Against Chemtrails and Geoengineering," an annual event that takes place in multiple cities worldwide. Organized by grassroots activists, the march aims to raise awareness about chemtrails and demand transparency and accountability from governments. Participants often carry signs, distribute flyers, and engage with passersby to share their concerns. These marches not only draw attention to the chemtrails theory but also create a sense of community among participants, reinforcing their commitment to the cause.

Community reactions to the chemtrails theory are not limited to activism but also include efforts to engage with policymakers and authorities. Activists frequently petition local and national governments to investigate and address their concerns about chemtrails. Public hearings and town hall meetings provide opportunities for citizens to voice their opinions and present their evidence to elected officials. While these efforts often face skepticism and resistance from authorities, they highlight the determination of chemtrails believers to seek recognition and action.

Despite the fervor of chemtrails activism, these efforts have also faced significant challenges and criticisms. Skeptics and scientific experts argue that the claims made by chemtrails activists lack credible evidence and are based on misunderstandings of atmospheric science. They caution that the spread of misinformation can divert attention and resources away from legitimate environmental and public health issues. Additionally, the confrontational nature of some activism efforts can create divisions within communities, straining relationships and fostering distrust.

In response to these challenges, some activists have sought to adopt a more collaborative and evidence-based approach. By engaging with scientists, environmental organizations, and policymakers, they aim to bridge the gap between their concerns and mainstream understanding. Efforts to promote open dialogue and mutual respect can help build trust and foster a more constructive discourse around environmental issues.

In conclusion, community reactions and activism have played a significant role in the spread and persistence of the chemtrails conspiracy theory. Through grassroots organizing, social media campaigns, public events, and engagement with policymakers, chemtrails activists have mobilized support and raised awareness about their concerns. While these efforts have faced challenges and criticisms, they underscore the importance of addressing public fears and promoting informed and evidence-based discussions. As we continue to explore the real-world impact of the chemtrails narrative, fostering open dialogue and collaboration between activists, scientists, and policymakers remains essential for addressing misinformation and promoting a healthier public discourse.

Media Coverage and Public Discourse

The evolution of media coverage surrounding chemtrails has played a pivotal role in shaping public discourse and influencing the persistence of the conspiracy theory. From initial sensationalist reports to more balanced and critical examinations, the way chemtrails have been presented in the media reflects broader trends in journalism and the public's appetite for conspiracy narratives.

In the early stages, media coverage of chemtrails was often sensationalist, with local news stations and tabloid newspapers running dramatic stories about mysterious trails in the sky and their purported health effects. These reports frequently featured interviews with concerned citizens and self-proclaimed experts who asserted that chemtrails were part of a covert operation. The lack of critical examination and the emphasis on anecdotal evidence gave the theory an air of legitimacy. Headlines like "Are We Being Sprayed?" and "Mystery in the Skies" captured public attention and fueled curiosity and concern.

As the theory gained traction, mainstream media outlets began to take notice, with some programs dedicating segments to exploring the claims. However, the balance between responsible journalism and sensationalism remained a challenge. While some investigative reports sought to debunk the theory by interviewing scientists and presenting evidence-based explanations, others perpetuated the narrative by giving equal weight to unverified claims. This "both-sides" approach, while intended to provide balanced coverage, often resulted in false equivalence, where fringe theories were presented alongside established scientific facts.

Over time, more comprehensive and critical coverage emerged. Major news organizations began to publish articles and documentaries that delved into the origins of the chemtrails theory, the psychological and sociological factors driving belief, and the scientific rebuttals. These pieces often highlighted the findings of environmental and atmospheric scientists, who explained that contrails are a natural byproduct of jet engine exhaust and are composed of water vapor and ice crystals. By presenting clear, evidence-based explanations, these reports aimed to dispel the myths and misinformation surrounding chemtrails.

Digital media has also played a significant role in shaping public discourse on chemtrails. Online platforms, including news websites, blogs, and social media, offer a space for both proponents and skeptics to share their views. The rapid dissemination of information online means that articles, videos, and social media posts can reach a wide audience quickly. This democratization of information has its benefits, but it also presents challenges in terms of content moderation and the spread of misinformation. While credible news outlets strive to maintain journalistic integrity, the sheer volume of content and the virality of sensationalist claims can make it difficult to control the narrative.

Case studies of key media events highlight the impact of coverage on the chemtrails narrative. For instance, the 2014 public hearing in Shasta County, California, which drew significant media attention, exemplified how local news coverage can influence public perception. The hearing featured emotional testimonies from residents and advocacy groups, as well as expert rebuttals from scientists. The media coverage of the event captured the tension between belief and skepticism, providing a microcosm of the broader discourse on chemtrails.

Another notable example is the increased scrutiny and debunking efforts by fact-checking organizations like Snopes and FactCheck.org. These organizations have published detailed articles addressing the chemtrails theory, explaining the science behind contrail formation, and debunking common misconceptions. Their work has contributed to a more informed public discourse, although the reach and influence of fact-checking can be limited in the face of entrenched beliefs.

In conclusion, media coverage has played a crucial role in shaping public discourse on chemtrails. From sensationalist reports to more balanced and critical examinations, the evolution of coverage reflects broader trends in journalism and the public's engagement with conspiracy theories. Understanding the role of media in influencing beliefs is essential for addressing the spread of misinformation. As we continue to explore the real-world impact of the chemtrails narrative, promoting responsible journalism, media literacy, and critical thinking remains key to fostering an informed and rational public discourse.

Chapter 11: The Future of the Chemtrails Theory

Current Trends and Developments

As we look to the future of the chemtrails theory, it's crucial to examine the current trends and developments within the community and broader public discourse. Despite ongoing scientific debunking, the theory remains persistent, fueled by a combination of technological advancements, media dynamics, and evolving social landscapes.

One of the most notable current trends is the continued evolution of digital platforms that facilitate the spread of chemtrails content. Social media remains a powerful tool for conspiracy theorists, enabling the rapid dissemination of information and providing a space for like-minded individuals to connect and reinforce their beliefs. Platforms like Facebook, YouTube, and Twitter are still heavily used by chemtrails proponents to share photos, videos, and personal anecdotes. These platforms' algorithms often favor engaging content, which can lead to the amplification of sensationalist posts about chemtrails, keeping the theory in public view.

Moreover, the rise of new technologies such as artificial intelligence and deepfake videos presents both opportunities and challenges for the future of chemtrails discourse. On one hand, AI can be used to develop more sophisticated methods for detecting and debunking misinformation. On the other hand, the same technology can be exploited by conspiracy theorists to create convincing fake evidence, making it harder for the public to distinguish between fact and fiction. This dynamic underscores the importance of continued vigilance and innovation in the fight against misinformation.

The ongoing global debate about climate change and environmental intervention also intersects with the chemtrails narrative. Geoengineering, the deliberate large-scale intervention in the Earth's natural systems to counteract climate change, is a topic of legitimate scientific research. However, it is frequently conflated with chemtrails by conspiracy theorists, who argue that such interventions are already happening in secret. As discussions around geoengineering become more prominent, we can expect the chemtrails theory to adapt and evolve in response, potentially gaining new adherents who are concerned about the implications of these technologies.

Within the chemtrails community itself, there are signs of both consolidation and fragmentation. Some proponents are doubling down on their beliefs, organizing more sophisticated campaigns and collaborating with other conspiracy theorist groups. This consolidation can lead to a more cohesive and resilient community, capable of sustaining the narrative despite external challenges. At the same time, there are internal divisions regarding the specifics of the theory, such as the exact nature and purpose of the alleged spraying. These debates can lead to fragmentation, with different factions promoting slightly different versions of the theory.

Another significant development is the increasing attention from mainstream media and fact-checking organizations. Major news outlets have published detailed investigations and documentaries that critically examine the chemtrails theory, providing evidence-based rebuttals. Fact-checking websites like Snopes and FactCheck.org continue to address chemtrails claims, helping to inform the public and counteract misinformation. These efforts contribute to a more informed public discourse, although their impact may be limited by the entrenched beliefs of some conspiracy theorists.

In conclusion, the future of the chemtrails theory will likely be shaped by a combination of technological advancements, media dynamics, and evolving public discourse around environmental intervention. While digital platforms continue to play a significant role in spreading the theory, new technologies and increased scrutiny from mainstream media and fact-checking organizations offer potential avenues for debunking and mitigating its impact. Understanding these current trends and developments is essential for anticipating the future trajectory of the chemtrails narrative and developing strategies to address misinformation and promote scientific literacy. As we continue to explore the evolution of the chemtrails theory, fostering a critical and informed public remains a key priority.

Potential for New Evidence or Debunking

The future trajectory of the chemtrails conspiracy theory may hinge significantly on the potential for new evidence, whether supporting or debunking the claims. Advancements in atmospheric science, environmental monitoring, and investigative methodologies offer both opportunities and challenges in addressing the chemtrails narrative. Exploring these possibilities is crucial for understanding how new findings might influence public belief and the spread of the theory.

One avenue for new evidence lies in the ongoing advancements in atmospheric science and environmental monitoring. The development of more sophisticated satellite technology and high-altitude atmospheric research aircraft allows scientists to gather detailed data on the composition and behavior of contrails. These advancements can provide clearer insights into the natural processes that create and sustain contrails, potentially offering definitive explanations for the phenomena often misinterpreted as chemtrails. Enhanced monitoring capabilities can also help identify and address any genuine environmental concerns more effectively, thereby countering the narrative that authorities are concealing information.

Furthermore, the role of governmental and independent investigations remains critical in addressing chemtrails claims. Over the years, various governmental agencies and independent researchers have conducted studies and released reports debunking the theory. Continued commitment to such investigations, coupled with transparent communication of findings, can help build public trust in credible sources. The involvement of international organizations and collaborative research efforts can also lend additional weight to debunking efforts, demonstrating a unified scientific consensus against the chemtrails theory.

In addition to scientific advancements, the role of citizen science and crowdsourced data collection can be influential. Engaging the public in scientific research related to atmospheric phenomena can foster a deeper understanding and appreciation of the scientific method. Initiatives that allow citizens to participate in data collection and analysis can demystify the processes behind contrail formation and provide firsthand evidence to counteract misinformation. By involving the public in scientific inquiry, these initiatives can empower individuals with knowledge and critical thinking skills, making them less susceptible to conspiracy theories.

However, the potential for new evidence supporting the chemtrails theory also exists, albeit limited. Should any credible evidence emerge that indicates the presence of harmful substances in contrails or deliberate atmospheric interventions, it would necessitate a thorough and transparent investigation. Addressing any valid environmental or health concerns is essential for maintaining public trust and ensuring that scientific research remains objective and unbiased. The scientific community must remain open to new findings while adhering to rigorous standards of evidence and peer review.

The impact of new evidence, whether supportive or debunking, on public belief in chemtrails will largely depend on how it is communicated and received. Effective science communication plays a pivotal role in shaping public perception. Clear, accessible, and transparent dissemination of scientific findings can help counteract misinformation and build public trust. Conversely, the presentation of new evidence in a sensationalist or opaque manner can fuel skepticism and conspiracy thinking.

In conclusion, the potential for new evidence to influence the future of the chemtrails conspiracy theory is substantial. Advancements in atmospheric science, environmental monitoring, and citizen science offer opportunities to provide definitive explanations and address genuine concerns. The role of governmental and independent investigations remains crucial in debunking the theory and maintaining public trust. Effective communication of scientific findings is essential for shaping public perception and countering misinformation. As we continue to explore the future of the chemtrails narrative, fostering a culture of scientific inquiry, transparency, and critical thinking is key to promoting an informed and rational public discourse.

The Ongoing Debate and its Implications for Society

The ongoing debate surrounding the chemtrails conspiracy theory is not merely an intellectual exercise; it has real and significant implications for society. The persistence of this theory reflects broader trends in public trust, information dissemination, and the societal impacts of conspiracy thinking. Understanding these implications is essential for addressing the challenges posed by the chemtrails narrative and fostering a more informed and rational public discourse.

The debate between proponents and skeptics of the chemtrails theory is marked by deeply entrenched positions on both sides. Proponents argue that the visible trails left by aircraft are evidence of a covert operation to manipulate the environment and public health. They cite anecdotal reports, selective interpretations of scientific studies, and perceived government secrecy as evidence supporting their claims. Skeptics, on the other hand, rely on established scientific explanations for contrails, extensive environmental monitoring data, and peer-reviewed research to debunk the theory. This polarization creates a challenging landscape for constructive dialogue, as each side often dismisses the other's evidence and motivations.

One of the primary implications of the ongoing chemtrails debate is its impact on public trust in science and governmental institutions. The belief in chemtrails is often rooted in a broader mistrust of authority and skepticism toward official narratives. This distrust can erode public confidence in scientific research, environmental monitoring, and health recommendations. When individuals perceive that authorities are concealing the truth or acting in their own interests, it becomes challenging to promote evidence-based policies and public health initiatives. The chemtrails theory exemplifies how conspiracy thinking can undermine efforts to address genuine environmental and health issues, diverting attention and resources away from legitimate concerns.

The societal implications of the chemtrails debate extend to the realm of information dissemination and media consumption. The rise of digital media and social platforms has transformed how information is shared and consumed, making it easier for misinformation to spread rapidly. The algorithms that prioritize engaging content can amplify sensationalist and conspiratorial claims, making them more visible and accessible to the public. This dynamic creates an environment where conspiracy theories can thrive, as individuals are exposed to information that reinforces their preexisting beliefs and biases. The challenge lies in promoting media literacy and critical thinking skills to help individuals navigate this complex information landscape and discern credible sources from unreliable ones.

The psychological and social implications of the chemtrails theory are also significant. Believers often experience heightened anxiety and fear, driven by the perception that they are being exposed to harmful chemicals. This sense of threat can have detrimental effects on mental health, contributing to stress, paranoia, and social isolation. Furthermore, the chemtrails narrative can create divisions within communities, straining relationships between believers and non-believers. These divisions can hinder constructive dialogue and collaboration, making it more difficult to address shared environmental and health concerns.

Despite these challenges, there are opportunities to address the implications of the chemtrails debate through education, public awareness, and open dialogue. Promoting scientific literacy and critical thinking in schools and communities can empower individuals with the skills to evaluate information critically and make informed decisions. Public awareness campaigns that provide clear and accessible explanations of scientific concepts can help counteract misinformation and build trust in credible sources. Additionally, fostering open dialogue between scientists, policymakers, and the public can create a more inclusive and transparent decision-making process, addressing concerns and building consensus.

In conclusion, the ongoing debate surrounding the chemtrails conspiracy theory has significant implications for society, impacting public trust, information dissemination, and psychological well-being. Understanding these implications is essential for developing strategies to address the challenges posed by the chemtrails narrative and promote a more informed and rational public discourse. As we continue to explore the future of the chemtrails theory, fostering a culture of scientific inquiry, transparency, and critical thinking remains key to mitigating the impact of conspiracy thinking and promoting a healthier society.

The Role of Education and Public Awareness

Education and public awareness play pivotal roles in addressing the chemtrails conspiracy theory and promoting a more scientifically literate society. By fostering critical thinking and providing accurate information, educational initiatives can counteract misinformation and help individuals make informed decisions. Developing strategies for effective education and public awareness is essential for mitigating the impact of conspiracy theories like chemtrails.

One of the most effective ways to combat misinformation is through integrating critical thinking and scientific literacy into school curriculums. From an early age, students should be taught how to evaluate sources, analyze evidence, and think critically about the information they encounter. Lessons on the scientific method, the peer-review process, and the principles of evidence-based reasoning can equip students with the skills needed to discern credible information from unreliable claims. By fostering a culture of inquiry and skepticism, educators can empower students to question and evaluate conspiracy theories critically.

Higher education institutions, such as universities and colleges, also have a significant role in promoting scientific literacy. Courses on media literacy, environmental science, and critical thinking can provide students with a deeper understanding of how misinformation spreads and how to counteract it. Additionally, universities can offer public lectures, seminars, and workshops on topics related to conspiracy theories and scientific skepticism, engaging the broader community and promoting informed public discourse.

Public awareness campaigns are another crucial tool for addressing the chemtrails theory. Government agencies, scientific organizations, and non-profit groups can collaborate to create campaigns that provide clear and accessible information about contrails, atmospheric science, and the debunking of chemtrails claims. These campaigns can utilize various media, including social media, websites, brochures, and public service announcements, to reach a wide audience. By presenting evidence-based information in an engaging and relatable manner, these campaigns can help counteract the influence of conspiracy theories.

Successful case studies demonstrate the effectiveness of education and public awareness in mitigating belief in conspiracy theories. For example, the "Science Literacy Week" initiative in Canada, organized by the Natural Sciences and Engineering Research Council (NSERC), promotes public understanding of science through events, workshops, and educational resources. By engaging the public in hands-on scientific activities and providing opportunities to interact with scientists, this initiative helps demystify science and build trust in credible sources.

Another example is the "Skeptics in the Pub" movement, which organizes informal gatherings where experts discuss various scientific topics, including conspiracy theories. These events provide a platform for open dialogue and critical examination of misinformation, fostering a sense of community and shared inquiry. By making scientific discussions accessible and enjoyable, initiatives like "Skeptics in the Pub" can reach individuals who might not otherwise engage with formal scientific education.

The role of media in promoting public awareness cannot be overstated. Journalists and media professionals have a responsibility to provide accurate and balanced reporting on scientific issues, avoiding sensationalism and false equivalence. Training programs for journalists on how to cover conspiracy theories and scientific topics can help ensure that media coverage is informative and evidence-based. Collaborations between media outlets and scientific organizations can also enhance the quality and credibility of information presented to the public.

In conclusion, education and public awareness are critical components in addressing the chemtrails conspiracy theory and promoting scientific literacy. By integrating critical thinking and scientific principles into education, creating engaging public awareness campaigns, and fostering open dialogue, we can empower individuals to evaluate information critically and make informed decisions. Understanding the role of education and public awareness in mitigating belief in conspiracy theories is essential for developing effective strategies to counteract misinformation and promote a more informed and rational society. As we continue to explore the future of the chemtrails narrative, fostering a culture of inquiry, skepticism, and evidence-based reasoning remains key to building trust in science and credible sources.

Future Directions for Research and Public Policy

Looking ahead, the future directions for research and public policy will play a crucial role in addressing the chemtrails conspiracy theory and promoting a scientifically literate society. Advancements in atmospheric science, environmental monitoring, and policy-making can help debunk the theory, address genuine environmental concerns, and foster public trust in credible sources.

One of the key areas for future research is the continued study of contrail formation and its impact on the environment. Scientists must delve deeper into understanding how contrails form under varying atmospheric conditions and their long-term effects on climate. Advanced satellite technology and high-altitude research aircraft can provide more precise data on the chemical composition of contrails and their interaction with natural cloud formations. By producing comprehensive and transparent research, the scientific community can offer clear explanations that counteract misinformation.

Public policy can also play a significant role in addressing the environmental concerns that fuel the chemtrails theory. Governments should invest in robust environmental monitoring systems that provide accurate data on air quality and atmospheric conditions. Transparent reporting of these findings can help build public trust and dispel fears about hidden environmental dangers. Policy measures that promote open access to environmental data and encourage collaboration between scientists, policymakers, and the public can ensure that credible information is widely available and easily understood.

Another important aspect of future research is the examination of the psychological and sociological factors driving belief in conspiracy theories like chemtrails. Understanding the cognitive biases, social dynamics, and emotional triggers that contribute to conspiracy thinking can inform the development of targeted educational initiatives and public awareness campaigns. Interdisciplinary research that combines insights from psychology, sociology, and communication studies can help create more effective strategies for countering misinformation and promoting critical thinking.

In terms of public policy, efforts to regulate and moderate online content are crucial. Social media platforms and online forums are breeding grounds for the spread of conspiracy theories, and policymakers must work with tech companies to address this issue. Implementing policies that encourage responsible content moderation, fact-checking, and the promotion of credible sources can help reduce the visibility of misinformation. However, these measures must balance the need to combat false information with the protection of free speech rights, ensuring that any regulations are transparent and fair.

Educational initiatives must also be a priority for future policy-making. Integrating critical thinking, media literacy, and scientific literacy into school curriculums can equip individuals with the skills needed to evaluate information critically. Public awareness campaigns that engage the broader community through workshops, lectures, and interactive events can further promote scientific understanding and skepticism toward conspiracy theories. By fostering a culture of inquiry and evidence-based reasoning, educational policies can help build a more informed and rational public.

Collaborative efforts between scientists, policymakers, and the public are essential for addressing the chemtrails theory and promoting a healthier public discourse. Forums for open dialogue and public participation in scientific research can bridge the gap between experts and the community, ensuring that concerns are addressed and misinformation is countered. Initiatives that involve citizens in environmental monitoring and data collection can empower individuals with firsthand experience and knowledge, reducing the appeal of conspiracy theories.

In conclusion, the future directions for research and public policy are critical for addressing the chemtrails conspiracy theory and fostering a scientifically literate society. Advancements in atmospheric science, transparent environmental monitoring, and interdisciplinary research can provide clear explanations and counteract misinformation. Public policy measures that promote responsible content moderation, critical thinking, and open dialogue can build public trust and reduce the impact of conspiracy thinking. As we continue to explore the future of the chemtrails narrative, fostering a culture of scientific inquiry, transparency, and collaboration remains key to promoting an informed and rational public discourse.

Chapter 12: Conclusion

Summary of Key Points Discussed

As we draw to a close, it's important to reflect on the key points and themes explored throughout this book on the chemtrails conspiracy theory. Our journey began with an examination of the origins and evolution of the chemtrails narrative, tracing its roots to the mid-1990s and following its growth through various phases of public interest and media coverage.

We delved into the scientific foundations of contrails and how they differ from the alleged chemtrails. The formation of contrails, resulting from the condensation of water vapor in aircraft exhaust, was explained in detail. Through scientific studies and expert opinions, we debunked the claims that these trails contain harmful chemicals deliberately sprayed for nefarious purposes. The peer-reviewed research consistently showed that contrails are composed of water vapor and ice crystals, with no evidence supporting the chemtrails theory.

The book also explored the psychological and sociological factors driving belief in chemtrails. We examined the cognitive biases, such as confirmation bias and pattern recognition, that make conspiracy theories appealing. The sense of control, community, and identity derived from believing in chemtrails was discussed, highlighting the emotional and cognitive satisfaction that sustains these beliefs. Sociological factors, including the role of social networks, media dynamics, and cultural contexts, were analyzed to understand how the chemtrails narrative spreads and persists.

In addition to the psychological and sociological perspectives, we investigated the real-world impact of the chemtrails theory. Notable incidents and investigations were examined, revealing how these events influenced public belief and the spread of the theory. The legal and political responses to chemtrails claims, including lawsuits, legislative measures, and government inquiries, were discussed to understand their outcomes and implications. We also looked at the impact on public health and the environment, with scientific research consistently finding no evidence to support the health and environmental concerns attributed to chemtrails.

Community reactions and activism were another critical area of exploration. We discussed how grassroots movements and public awareness campaigns have mobilized citizens and drawn attention to the chemtrails theory. The methods and strategies used by activists, as well as the social dynamics within these communities, were analyzed to understand their influence on public opinion.

Media coverage and public discourse played a significant role in shaping the chemtrails narrative. The evolution of media representation, from sensationalist reports to more balanced and critical examinations, was traced to understand its impact on public perception. The role of alternative media and citizen journalism in spreading the theory was also examined, highlighting the challenges and opportunities for addressing misinformation.

Finally, we considered the future of the chemtrails theory, exploring current trends and developments, the potential for new evidence or debunking, and the ongoing debate's implications for society. The importance of education and public awareness in addressing the theory was emphasized, with strategies for promoting scientific literacy and critical thinking discussed. Future directions for research and public policy were outlined, underscoring the need for collaboration between scientists, policymakers, and the public.

In summary, this book has provided a comprehensive examination of the chemtrails conspiracy theory, addressing its origins, scientific rebuttals, psychological and sociological factors, real-world impact, media influence, and future directions. By understanding these elements, we can better address the spread of misinformation and promote a more informed and rational public discourse. As we look to the future, fostering a culture of critical thinking, scientific literacy, and open dialogue remains essential for countering conspiracy theories and building a healthier society.

Final Thoughts on the Chemtrails Conspiracy Theory

The persistence and adaptability of the chemtrails conspiracy theory reveal much about the nature of conspiracy beliefs and their broader implications for society. Despite substantial scientific evidence debunking the theory, the narrative of chemtrails continues to captivate a segment of the population, driven by deep-seated mistrust of authority, cognitive biases, and the human need for simple explanations to complex phenomena.

One of the key reasons for the enduring appeal of the chemtrails theory is its adaptability. The narrative has evolved over time, incorporating new elements and responding to shifts in public concern. Originally focused on alleged weather modification, the theory has expanded to include claims about population control, mind control, and various environmental and health impacts. This flexibility allows the theory to remain relevant and appealing, even as specific claims are debunked.

The broader implications of the chemtrails theory for society are significant. Conspiracy theories like chemtrails can undermine public trust in scientific institutions and governmental authorities. When large segments of the population believe that they are being misled or manipulated, it becomes challenging to build consensus on critical issues such as public health, environmental protection, and climate change. The erosion of trust in credible institutions can lead to a fragmented and polarized society, where misinformation flourishes, and evidence-based policy-making becomes difficult.

Addressing and debunking conspiracy theories like chemtrails is not without its challenges. One of the primary difficulties lies in the deeply ingrained nature of these beliefs. Cognitive biases, such as confirmation bias and the Dunning-Kruger effect, can make individuals more resistant to evidence that contradicts their views. Efforts to debunk conspiracy theories can sometimes backfire, reinforcing the beliefs of those who feel their views are being attacked or dismissed. This dynamic underscores the importance of approaching the issue with empathy, respect, and a focus on education and critical thinking.

Critical thinking and scientific literacy are essential tools in countering conspiracy theories. By fostering a culture of inquiry and skepticism, we can equip individuals with the skills needed to evaluate information critically and make informed decisions. This involves not only teaching the principles of the scientific method and evidence-based reasoning but also encouraging open dialogue and the questioning of assumptions. Promoting critical thinking is not about dismissing alternative viewpoints but about ensuring that beliefs are grounded in credible evidence and rigorous analysis.

The ongoing relevance of the chemtrails theory in public discourse highlights the need for continued efforts to address misinformation and promote a more informed society. As technology and media continue to evolve, new challenges and opportunities will arise in the fight against misinformation. Collaboration between scientists, educators, policymakers, and the public is essential for developing effective strategies to counteract conspiracy theories and build trust in credible sources.

In conclusion, the chemtrails conspiracy theory serves as a case study in the persistence and adaptability of conspiracy beliefs. Its implications for public trust, societal cohesion, and evidence-based policy-making are profound. By promoting critical thinking, scientific literacy, and open dialogue, we can address the challenges posed by conspiracy theories and work toward a more informed and rational society. The journey to counteract misinformation is ongoing, but with collective effort and commitment, we can make significant strides toward fostering a culture of inquiry and evidence-based reasoning. As we look to the future, the lessons learned from the chemtrails narrative will continue to inform our approaches to addressing and mitigating the impact of conspiracy theories.

Encouragement for Readers to Think Critically and Seek the Truth

In an age where misinformation can spread rapidly and widely, critical thinking and the pursuit of truth have never been more important. The chemtrails conspiracy theory, like many others, thrives on the absence of critical inquiry and the appeal of simple explanations to complex issues. As readers, cultivating the habit of thinking critically and seeking the truth is essential not only in debunking such theories but also in fostering a healthier, more informed society.

Critical thinking involves questioning assumptions, evaluating evidence, and considering alternative perspectives before forming conclusions. It's a skill that can be developed and honed through practice and education. By applying critical thinking to the information we encounter, we can avoid the pitfalls of cognitive biases and ensure that our beliefs are based on solid evidence rather than anecdote or speculation. This approach is particularly important when dealing with conspiracy theories, which often rely on emotional appeal and selective interpretation of facts.

To think critically, it is crucial to seek out credible sources and evidence-based explanations. This means relying on information from reputable scientific organizations, peer-reviewed journals, and experts in the field. It also involves being cautious of sources that lack transparency, fail to provide evidence, or use sensationalist language to provoke fear and mistrust. By prioritizing credible sources, we can build a foundation of knowledge that is grounded in reality and resistant to misinformation.

Questioning assumptions is another vital component of critical thinking. Conspiracy theories often thrive on unchallenged assumptions and logical fallacies. By scrutinizing the premises of a theory and asking whether they hold up under scrutiny, we can identify and address any weaknesses in the argument. This process of questioning should be applied not only to external information but also to our own beliefs and biases. Being open to changing our minds in the face of new evidence is a hallmark of true critical inquiry.

It's also important to be open to new information and alternative viewpoints. In a world where information is abundant and diverse, we must be willing to consider perspectives that differ from our own. Engaging with a variety of sources and viewpoints can help us develop a more nuanced understanding of complex issues and avoid the echo chamber effect, where we only hear information that reinforces our existing beliefs. This openness to new information is essential for personal growth and for fostering a more inclusive and informed public discourse.

Fostering a skeptical and inquisitive mindset can be achieved through various means. Reading widely, participating in discussions and debates, and seeking out educational opportunities are all ways to develop and refine critical thinking skills. Engaging with science communication platforms, attending public lectures, and following reputable science communicators on social media can provide valuable insights and keep us informed about the latest scientific developments.

As readers, we have a collective responsibility to engage in informed and rational discourse. By thinking critically and seeking the truth, we can contribute to a society that values evidence-based reasoning and is resilient to misinformation. This commitment to critical inquiry not only helps us navigate the complexities of the modern information landscape but also empowers us to make better decisions for ourselves and our communities.

In conclusion, the journey of debunking the chemtrails conspiracy theory and other forms of misinformation begins with a commitment to critical thinking and the pursuit of truth. By questioning assumptions, seeking out credible sources, and being open to new information, we can build a foundation of knowledge that is grounded in reality. As we continue to engage with the world around us, let us strive to foster a culture of inquiry, skepticism, and evidence-based reasoning, contributing to a more informed and rational society.

The Role of Individuals and Communities in Addressing Conspiracy Theories

As we grapple with the persistence of conspiracy theories like chemtrails, the role of individuals and communities becomes increasingly crucial. Collective responsibility in countering misinformation is not only a societal necessity but also an empowering endeavor. By promoting education, fostering public awareness, and building community support, we can combat the spread of false information and nurture a more informed populace.

Education remains a cornerstone in the fight against conspiracy theories. However, it is not enough to rely solely on formal education systems. Individuals must take an active role in their own learning, seeking out credible information and engaging with diverse viewpoints. Parents and guardians can instill critical thinking skills in their children from an early age, encouraging curiosity and skepticism. Schools and educators should emphasize the importance of scientific literacy and provide students with the tools to critically evaluate the information they encounter. By prioritizing these skills, we can prepare future generations to navigate the complexities of the information age with discernment and reason.

Public awareness campaigns are another vital component in addressing conspiracy theories. Government agencies, non-profit organizations, and community groups can collaborate to create and disseminate campaigns that educate the public about the nature of conspiracy theories and the importance of evidence-based reasoning. These campaigns should be accessible, engaging, and relatable, leveraging various media platforms to reach a broad audience. By raising awareness and providing accurate information, these initiatives can help counteract the allure of conspiracy narratives and build trust in credible sources.

Community support is essential for fostering a culture of inquiry and evidence-based reasoning. Local organizations, libraries, and community centers can host events such as workshops, lectures, and discussion groups focused on critical thinking and scientific literacy. These gatherings provide opportunities for individuals to engage with experts, ask questions, and share their concerns in a supportive environment. By creating spaces for open dialogue and mutual learning, communities can empower their members to challenge misinformation and seek out the truth.

Individuals also play a crucial role in countering misinformation within their social networks. Conversations with friends, family members, and colleagues can be powerful tools for addressing conspiracy beliefs. However, these discussions must be approached with empathy and respect. Instead of dismissing or ridiculing someone's beliefs, individuals should engage in thoughtful dialogue, asking questions and providing evidence-based information. Building trust and understanding is key to opening minds and fostering a willingness to reconsider entrenched beliefs.

Grassroots efforts and community initiatives can have a significant impact on combating conspiracy theories. Citizen science projects, where members of the public participate in scientific research, can demystify the scientific process and provide firsthand experience with evidence-based inquiry. Examples include community air quality monitoring programs or local climate observation initiatives. By involving individuals in the process of data collection and analysis, these projects can enhance scientific literacy and dispel misconceptions.

In conclusion, the role of individuals and communities in addressing conspiracy theories like chemtrails is both crucial and multifaceted. Through education, public awareness campaigns, community support, and grassroots initiatives, we can collectively counteract misinformation and promote a more informed society. By fostering a culture of critical thinking and evidence-based reasoning, we empower individuals to navigate the information landscape with discernment and build trust in credible sources. As we look to the future, the collective efforts of individuals and communities will be instrumental in addressing the challenges posed by conspiracy theories and promoting a healthier public discourse.

Looking Ahead to a More Informed and Rational Society

Envisioning a future where scientific literacy and critical thinking are widely embraced offers a hopeful prospect in the ongoing battle against misinformation and conspiracy theories like chemtrails. By fostering a culture of inquiry and evidence-based reasoning, we can work towards a society where individuals are better equipped to navigate complex information landscapes and make informed decisions.

In this vision, education serves as the bedrock for promoting scientific literacy. Schools and universities play a pivotal role in instilling critical thinking skills and a deep understanding of the scientific method from an early age. Curriculums that emphasize inquiry-based learning, where students engage in hands-on experiments and scientific investigation, can ignite curiosity and foster a lifelong appreciation for science. Educators are empowered to create learning environments that encourage questioning and exploration, helping students develop the skills needed to evaluate information critically.

The potential for new technologies and research to address misinformation is immense. Advances in artificial intelligence and machine learning can enhance the capabilities of fact-checking organizations and content moderation systems. These technologies can help identify and flag false information more effectively, providing users with reliable and accurate content. Additionally, ongoing research in cognitive science and psychology can offer deeper insights into the mechanisms of belief formation and the spread of misinformation, informing the development of targeted interventions and educational strategies.

Collaboration between scientists, policymakers, and the public is essential for fostering a more informed society. Scientists must continue to communicate their findings transparently and accessibly, engaging with the public through various media and outreach initiatives. Policymakers have a responsibility to support evidence-based decision-making and promote policies that enhance scientific literacy and public trust. Public engagement in scientific research, through citizen science projects and community initiatives, can bridge the gap between experts and the community, creating a more inclusive and participatory approach to addressing societal challenges.

As we reflect on the progress made and the work still to be done, it is important to acknowledge the collective efforts of individuals and organizations dedicated to promoting scientific literacy and countering misinformation. Grassroots movements, educational initiatives, and public awareness campaigns have made significant strides in raising awareness and fostering critical thinking. However, the journey is ongoing, and continued effort is needed to build on these successes and address emerging challenges.

Looking ahead, the hope for a more informed and rational society is anchored in our commitment to education, collaboration, and evidence-based reasoning. By fostering a culture that values inquiry and skepticism, we can empower individuals to make informed decisions and contribute to a healthier public discourse. The lessons learned from the chemtrails narrative serve as a reminder of the importance of vigilance and critical thinking in the face of misinformation.

In conclusion, the vision for a more informed and rational society is both achievable and essential. Through education, technological innovation, and collaborative efforts, we can counteract the spread of conspiracy theories and promote a culture of evidence-based reasoning. As we continue to address the challenges posed by misinformation, fostering scientific literacy and critical thinking remains at the heart of our efforts. Together, we can build a society where individuals are empowered to seek the truth, question assumptions, and engage in informed and rational discourse, paving the way for a brighter and more enlightened future.